MUSHROOMS AND TOADSTOOLS

WARD LOCK LIMITED

A KINGFISHER BOOK

First published in 1980 by Kingfisher Books Limited
(A Grisewood & Dempsey, Ward Lock joint company)
116 Baker Street, London W1M 2BB

© Kingfisher Books Limited 1980

Colour separations by Newsele Litho Ltd, Milan, London
Printed and bound in Italy by Vallardi Industrie Grafiche, Milan

BRITISH LIBRARY CATALOGUING
IN PUBLICATION DATA
Reid, Derek.
  Mushrooms and toadstools.—(Kingfisher guides).
  1. Fungi—Europe—Identification
  II Title II. Series
  589'.2'094   QK606.5
  ISBN 0-7063-5999-2

*The author wishes to thank Ronald Rayner
for his help in the preparation of this book.*

AUTHOR
DEREK REID

ILLUSTRATOR
BERNARD ROBINSON

DESIGNED AND EDITED
BY
JULIA KIRK

CONTENTS

INTRODUCTION

The term 'mushroom' was, until relatively recently, restricted to the field mushroom and horse mushroom (*Agaricus campestris* and *Agaricus arvensis* respectively) among the wild species, but included the cultivated mushroom (*Agaricus bisporus*). All other gill-bearing fungi, sometimes referred to as the agarics, were known as toadstools. However, under the influence of American usage, the word mushroom has been broadened to cover not only all the agarics but all the larger, fleshy fungi.

Fungi have a very varied appearance, ranging from patches on wood to brackets, coral-like tufts, simple clubs, rosettes, cauliflower-like structures or centrally or laterally stalked fruitbodies. In addition, the fertile surface may be smooth or have spines, pores or gills. The texture varies from woody, leathery or fleshy to gelatinous and the surface may be either smooth, velvety, hairy or scaly and either dry or glutinous.

Whatever their appearance, the fungi have certain features which clearly separate them from all other groups of plants. Apart from differing in shape, they all lack the green pigment chlorophyll and are unable to manufacture their own carbohydrates by the process of photosynthesis. Another distinction involves structure: the fungi are all formed of branching threads or hyphae. These threads are quite different from the cells which form the basic structure of other plants. Again, reproduction of the fungi is different from that of all other plant groups. Fungi spread by minute spores, visible only under a microscope. These are borne on or formed inside characteristic types of organ.

The Function of a Mushroom

When considering a mushroom, one must remember that it is only the reproductive fruitbody of a particular fungus and that it is produced on and nourished by an extensive network of hyphae. Most fungi obtain their food by the breakdown of plant or animal material and grow where this is most abundant. They are, therefore, frequent in woodland where there is an abundance of humus but they also grow in pastures, on heaths, sand-dunes, or mountaintops on wood, soil or dung. In such species, the network consists of a cobwebby mat of hyphae which branch out amongst the rotting vegetation on the woodland floor or amongst dead or decaying grass. This mat of hyphae is known as a mycelium. How-ever, some fungi are parasitic and certain of the larger fleshy fungi even parasitize other fleshy fungi. In these species, the mycelium is found to be concealed within the host tissue.

A fairy-ring formed by Agaricus arvensis, the Horse Mushroom.

Fairy-Rings

Fairy-ring formations occur when a mycelium grows out equally in all directions from a central point, advancing slowly on a circular front but drying off behind as food supplies are exhausted. The advancing hyphae secrete enzymes which break down humus in the soil into nitrites and nitrates. These substances act as fertilizers to the grass in the immediate vicinity, stimulating a zone of lush, bright green growth. Within this zone there is a narrow region of bare earth or stunted grass. This is due to the filaments clogging the air spaces in the soil so preventing penetration by water. So although it is here that the fruitbodies are produced, this zone is virtually suffering from drought. Within this region there is a further zone of lush green grass resulting from the death of the old exhausted filaments which are then broken down into nitrites and nitrates by bacteria, once again stimulating the grass to lush growth. So a typical fairy-ring consists of two zones of lush green grass separated by a zone of stunted growth. The commonest gill-bearing fungus to form fairy-rings on lawns is the 'Fairy-Ring Champignon' (*Marasmius oreades*) but many other species will do so, including the field mushroom (*Agaricus campestris*).

Development of the Fruitbody

Once a mycelium is established, it continues to grow and eventually builds up sufficient food reserves to enable fruitbody formation to occur. For example, in the case of an *Amanita* species, dense knots of filaments form at certain points and each of these eventually develops into a 'button-stage' fruitbody. If this were to be cut through vertically, the

9

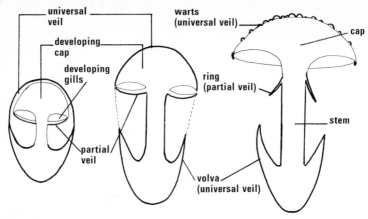

DEVELOPMENT OF THE FRUITBODY

button would be seen to consist of a tiny cap with gills on the underside and a stalk. The section would show that the gills are entirely covered by a membrane stretching from near the stem apex to the margin of the young cap – the 'partial veil'. It would also show that the entire developing fruitbody is completely enclosed within another membrane – the 'universal veil'.

As the young fruitbody enlarges, the stalk elongates, pushing the cap upwards until the pressure causes the universal veil to break. If the surface of the cap is sticky or if the membrane is fibrous, the cap slips through cleanly and the universal veil is left at the base of the stem as a sac-like sheath – the volva. If the membrane has a delicate powdery texture, part of it sticks to the cap surface as a mealy covering and the remainder forms a volva at the base of the stem, although reduced to one or more zones of scales which may be quite inconspicuous.

The stem continues to grow, accompanied by an increase in the size of the cap. As the cap expands, the remnants of the universal veil become stretched and break up into warts or mealy scales which become more and more dispersed toward the cap margin. At the same time the partial veil protecting the developing gills also becomes stretched and finally breaks away from the cap margin to fall back on the stem as a ring (annulus). In some genera there may be a partial veil but no universal veil, for example, *Lepiota* species, so that the mature fruitbodies have a ring but no volva. The partial veil may be membranous as in *Lepiota* or it may be cobwebby as in *Cortinarius*, where it is referred to as a 'cortina'. When the cortina breaks, it collapses onto the stem leaving a zone of fibrils, often made more obvious by trapped spores. In yet other genera, there is neither a partial nor a universal veil and here the fruitbodies lack both ring and volva.

Reproduction

The spores, which are the equivalent of seeds in flowering plants, are produced on the surface of the gills on structures known as basidia. In the bracket fungi these structures line the inside of the tubes, the mouths of which form the pore-bearing undersurface. In the hedgehog fungi basidia are found over the spines and in the stereoid fungi they are produced on the smooth underside of the brackets. Spores are of constant size, shape and colour for every species and spore characters are extremely important in the identification of fungi.

At maturity the spores are shot away for a short distance and then fall free of the gills or pores under their own weight, to be dispersed by air currents. On reaching a suitable habitat, the spores germinate, each producing a new mycelium, which must fuse with another mycelium of the same species in order to form fruitbodies. Fruitbody formation can then occur provided the mycelium has sufficient food material and that the temperature, humidity and light requirements are met.

The fact that there has to be fusion of two mycelia before fruitbodies can be produced explains why it is that the spread of exotic species from one continent to another occurs so seldom under natural conditions. To do so, spores of a certain species would have to be deposited by the wind within a centimetre or so of each other, both would have to germinate, fusion would have to occur and finally climatic conditions would have to be exactly right. When introduction of foreign species does occur, it is more likely to be in soil or on timber where the fused mycelia are already established.

Fungi in Commerce

Fungi are of considerable economic importance. A number of species are parasitic on other plants and cause considerable losses to foresters, farmers and gardeners alike. *Heterobasidion annosum*, a bracket fungus, is responsible for one of the most serious diseases of conifers in this country and *Armillariella mellea* (Honey fungus) causes the death of many deciduous trees and conifers each year. The latter fungus is not only a problem to foresters but also to ordinary gardeners, for it frequently makes it impossible to grow woody plants in small gardens if heavy infection becomes established.

Balancing the harmful effects of parasitic fungi in forestry are other species, which form mycorrhizae or 'fungus roots' with the roots of trees. Such roots function more efficiently in absorbing food materials from the soil, and trees grow better and more rapidly. The filaments of the fungus penetrate between the cells of the roots without harming them. Food materials absorbed by the filaments from the soil become available to the tree while various substances manufactured by the tree become available to the fungus. The relationship is therefore mutually beneficial. It also has commercial applications, for when trying to establish plantations of trees in new areas or in a foreign country, it is often necessary to ensure that the region is infected with the correct fungus if the trees are to make headway. The relationship also explains why some fungi are associated with particular hosts, for example, *Amanita muscaria* with birch.

Fungi as Food

There are many 'old wives' tales' concerning the distinction between edible and poisonous mushrooms. For example, if the cap peels, or if it does not turn a silver spoon black during cooking then it is safe to eat. These and many other so-called tests for edibility are all unreliable and should be disregarded. The only safe way is to go out with someone who can point out the edible species and the features by which they are recognized. In general, you should always avoid any white-spored species with both a ring and a sac-like volva. If you think you have an edible field or horse mushroom (*Agaricus campestris* or *Agaricus arvensis*) you should ensure that the spore-print is purplish-black and that there is a ring on the stem. Also, scratch the base of the stem and the cap and avoid any fruitbodies which bruise bright yellow.

After collection, fungi should be carried in shallow open baskets, each collection being placed in a paper bag, or if the species is very small or delicate, in tins lightly packed with moss to prevent damage. Plastic bags should be avoided, as condensation frequently occurs and excess humidity causes rapid decay. Specimens should always be dug up with a knife to ensure that any volval remains at the base of the stem are present. Finally, particular attention should be paid to the tree under or on which the fungus is growing, since this is an important feature in identification.

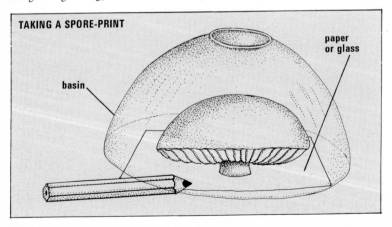

TAKING A SPORE-PRINT

paper or glass

basin

GILL ATTACHMENT AND FRUITBODY SHAPE

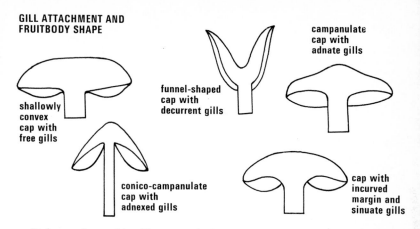

shallowly convex cap with free gills

funnel-shaped cap with decurrent gills

campanulate cap with adnate gills

conico-campanulate cap with adnexed gills

cap with incurved margin and sinuate gills

Before trying to identify an agaric (gill-bearing fungus), it is essential to take a spore-print to check the colour of the spore powder. To do this, cut off the stalk as near the cap as possible and place the cap (gills downward) on paper, or preferably glass, and cover with a basin or jar for 12 to 24 hours. If the gills are white the cap is best put on black paper but if the gills are coloured white paper should be selected. An exact replica of the gill arrangement, formed by the deposit of spores, will result. It is often necessary to support the basin on a pencil at one side to avoid undue humidity which might interfere with spore deposition. Spore-prints of very delicate or tiny species should be taken in tins to prevent drying out, but even here it is often necessary to leave the lid not quite closed. Spore deposits which may be white (sometimes pale-cream or faintly pink), brownish-pink, brown, purplish or black can be fixed by spraying with hair lacquer. When the colour of the spore-print has been established, you can then turn to that section of the book where the appropriate group of fungi with that colour of spore-print is dealt with.

When trying to assign an agaric to its correct genus it is necessary to note the type of gill attachment to the stem, and for which there are several terms. *Free* is self-explanatory; *decurrent* is when the gills run down the stem – the usual situation in many funnel-shaped fruitbodies; *adnate* is when the gill is attached by its entire width; *adnexed* when attachment is by a part only of the total width; *sinuate* when there is a notch in the gill where it joins the stem.

Inevitably, readers will find numerous species which are not described in this book, but in general most common species have been included. The author has tried to select only those species which can be identified with confidence without recourse to a microscope.

AGARICS

The agarics are those fleshy fungi which bear gills. They grow on the ground in pasture or woodland, on dung, trees or on woody debris. The stem may or may not bear a ring or conspicuous zone of fibrils and may or may not show a sheathing, sac-like structure at its base.

AMANITA

A genus of approximately 25 species, having in common a white spore-print, a volva at the base of the stem and usually, but not always, a membranous ring. The volva may be conspicuous and sac-like or reduced to a narrow rim or to concentric rings of scales. The cap is often ornamented with conspicuous warts, flat patches of tissue or hoary scales but when present these are superficial and are easily removed. The genus *Amanita* includes species which are amongst the most poisonous known agarics.

AMANITA PHALLOIDES
Death Cap

Cap: 6-9 cm diam., convex then shallowly convex; varying in colour from olive-green at the centre to yellowish-green nearer the smooth margin. The cap surface appears indistinctly radially streaky; usually naked, without trace of warts or scales.
Stem: 7-9 cm high; 1-1·5 cm wide; white, cylindrical or narrowed above and sheathed below in a conspicuous free-standing, white, sac-like volva.
Ring: near apex of stem; white.
Gills: white; virtually free.
Smell: when old unpleasant, cheesy.
Spore-print: white.
Habitat: deciduous woodland, especially with beech and oak. Autumnal. Fairly common.
POISONOUS, often fatal.

The streaky greenish cap, the sac-like volva, presence of a ring and the white spore-print are diagnostic of this deadly fungus.

Amanita
phalloides

Amanita
citrina

14

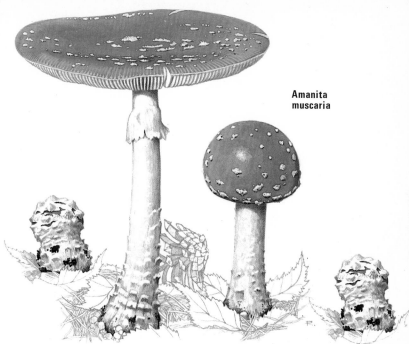

Amanita muscaria

AMANITA CITRINA
False Death Cap
Syn: *A. mappa*
Cap: 6-8 cm diam.; convex then flat; pale cream to pale lemon-yellow ornamented with a few large, or several smaller, thick, flat, whitish to brownish patches of velar tissue.
Stem: 8-11 cm high, 1-2 cm wide, tall in proportion to cap; white, cylindrical with a conspicuous basal bulb, up to 3 cm wide, with margin indicated by a prominent rim representing the volva.
Ring: near apex; white.
Gills: white to very pale cream.
Smell: of raw potato.
Spore-print: white.
Habitat: deciduous and coniferous woodland. Autumnal. Very common.

Formerly confused with the Death Cap and regarded as poisonous, but in reality it is harmless. *A. citrina* is easily separated from *A. phalloides* by cap-colour, and lack of a sac-like volva.

AMANITA MUSCARIA
Fly Agaric
Cap: up to 15 cm diam., convex, flattened to saucer-shaped, scarlet ornamented with white warts which become more dispersed toward the striate margin. With age the warts gradually disappear and may be completely lacking. The cap colour often fades to reddish-orange. In the button stage the cap is covered by thick white velar tissue which eventually cracks to form the scales and expose more and more of the red surface as expansion occurs.
Stem: up to 20 cm high, 3 cm wide, white to pale-yellowish above, cylindrical and brittle with a slightly broader base surmounted by a series of concentric zones of white scales representing the volva.
Ring: near apex of stem, white to very pale-yellowish.
Gills: white, almost free.
Habitat: Under birch, occasionally with pine and other trees. Autumnal. Common. POISONOUS but seldom lethal.

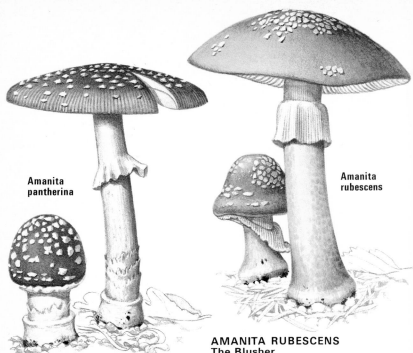

Amanita pantherina

Amanita rubescens

AMANITA PANTHERINA

Cap: 6-8 cm diam., convex, then flat, dark greyish-brown to olive-brown, sometimes paler and more yellowish-brown, ornamented with numerous, uniformly distributed small white pyramidal warts; margin somewhat striate.

Stem: 7-10 cm high, 1-1·5 cm wide, white, rather tall, cylindrical, slightly enlarged below where the volva disrupts into concentric rings, the uppermost forming a narrow, close-fitting but free collar or ridge.

Ring: near middle of stem, white.

Gills: white, free.

Flesh: white becoming orange-yellow under cap cuticle with potassium hydroxide.

Spore-print: white.

Habitat: deciduous woodland. Autumnal. Rare. POISONOUS.

Recognized by the brown cap ornamented with numerous small contrasting white pyramidal scales, the striate cap-margin, and collar-like formation of the volva.

AMANITA RUBESCENS
The Blusher

Cap: 8-12 cm diam., at first strongly bell-shaped, then convex, finally flat to shallowly saucer-shaped, varying in colour from pale pinkish-brown with a darker red-brown centre to entirely red-brown, ornamented with thin mealy or hoary patches of whitish, greyish or pale volval tissue which tend to disappear with age.

Stem: 8-11 cm high, 2·5-3·5 cm wide, stout and stocky, cylindrical with somewhat enlarged base, whitish becoming pale pinkish-brown below, especially when handled. Volva reduced to inconspicuous rows of concentric scales.

Ring: near apex of stem; white.

Gills: white, often spotted red with age.

Flesh: white becoming pinkish when cut and in insect holes.

Spore-print: white.

Habitat: deciduous and coniferous woodland. Late summer and autumn. One of the first mushrooms to fruit. Very common. Edible, but best avoided owing to possible confusion with poisonous species.

AMANITA EXCELSA
Syn. *A. spissa*

Cap: 9-12 cm, convex, flat, then saucer-shaped, greyish to umber-brown ornamented with thin, greyish mealy to hoary patches of velar tissue which may eventually disappear; margin smooth.

Stem: 8-10 cm high, 2-2·5 cm wide, often rather stocky, white, cylindrical, base bulbous with concentric rings of scales representing the volva, uppermost scales occasionally forming a distinct ridge.

Ring: near apex of stem, white.

Gills: white.

Spore-print: white.

Habitat: deciduous and coniferous woodland. Autumnal. Fairly common.

A. pantherina is similar but differs in the white pyramidal warts on the cap (which has a striate margin) and the ridge-like rim of volval warts at the base of the stem. *A. rubescens* differs in colour and in the reddening of the stem both internally and externally.

AMANITA FULVA
Tawny Grisette
Syn. *Amanitopsis fulva*

Cap: 3·5-5 cm diam., long remaining acorn-shaped, then bell-shaped, finally flat but often with central boss, bright tawny-brown, often darker at centre, naked. Margin striate, fluted or grooved.

Stem: up to 11 cm high, 1 cm wide, tall, fragile, hollow, cylindrical, pale-tawny with well-developed, sac-shaped volva, same colour as stem, and which often clings to the slightly enlarged base.

Ring: absent.

Gills: white, free.

Spore-print: white.

Habitat: coniferous and deciduous woodland. From late summer to autumn; one of the earliest mushrooms to appear. Very common. Edible.

A. crocea, a closely related but much rarer species, is more robust with a brighter orange-brown cap, and with a tendency for the entire stem surface to disrupt into coarse scales.

Amanita excelsa

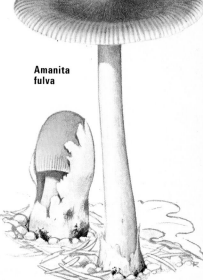

Amanita fulva

LEPIOTA

A genus of about 60 fungi, mostly rather small, delicate species with caps 1-5 cm diam., but a few, often placed in the genus *Macrolepiota*, are amongst our largest mushrooms. Most species have a scaly cap formed by the disruption of the surface. Most have a well-defined ring or ring zone, but this is sometimes deciduous, and only visible on young specimens. All species lack a volva, although there may be conspicuous scales at the base of the stem. The gills are free. Spore-print white. Most of the small species are toxic.

LEPIOTA CRISTATA

Cap: 2-4 cm diam., broadly campanulate, surface disrupting into tiny red-brown scales on a white background; the scales rapidly disappear except around a small, central, similarly coloured disc.

Stem: 2-5 cm high, 4-6 mm wide, white, cylindrical to slightly enlarged below.

Ring: near apex of stem, white, membranous, deciduous.

Gills: white, free.

Smell: unpleasant, sour, reminiscent of Earth Balls (*Scleroderma spp*).

Spore-print: white.

Habitat: deciduous woodland, often in grass along rides. Autumnal. Common. POISONOUS.

Easily confused with other small *Lepiota* species, but when examined under a strong lens most of these show minute pyramidal tufts of hairs over the central disc of the cap: these are lacking in *L. cristata*.

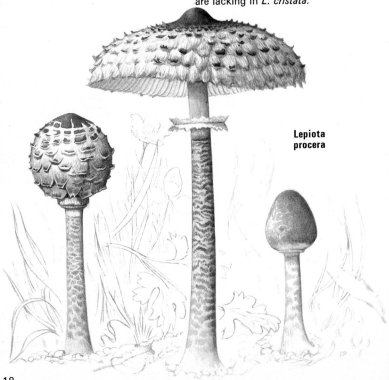

Lepiota procera

18

LEPIOTA PROCERA
Parasol Mushroom
Cap: 12-22 cm diam., umbrella-shaped with nipple-like boss at centre, surface disrupting into large, often upturned, dark-brown scales on a dirty white, coarsely fibrillose background; scales larger and widely dispersed towards margin, smaller, more densely arranged at centre with nipple uniformly dark-brown.

Stem: 20-26 cm high, 1·5-2·0 cm wide, tall, narrow, cylindrical with bulbous base up to 4 cm wide, pale, ornamented with dark-brown, zig-zag markings.

Ring: large, spreading with double edge, whitish, movable.

Gills: soft to touch, free, attached to collar-like rim surrounding the stem apex.

Flesh: white.

Spore-print: white.

Habitat: pastures, edge of woods grassy rides. Aumtunal, Occasional. Edible.

Unmistakable, distinguished by habitat, dark scaly cap, snake-like markings on stem, movable ring and unchanging white flesh. The movable ring is a feature of the large *Lepiota* species. Another feature is that the cap is easily separated from the stem due to a 'ball and socket' attachment.

LEPIOTA RHACODES
Shaggy Parasol
Cap: 8-12 cm diam., convex, lacking a central boss, surface disrupting into yellowish-brown, upturned shaggy scales on a dirty-white fibrillose background, except at the centre which is uniformly coloured and often darker.

Stem: 12-15 cm high, 1·5-2·0 cm wide, whitish, cylindrical with bulbous base up to 3 cm wide.

Ring: large, spreading, whitish, movable.

Gills: deep, soft to touch, free, attached to collar-like rim surrounding stem apex.

Flesh: white, reddening in stem when cut.

Spore-print: white.

Habitat: deciduous and coniferous woodland and shrubberies. Autumnal. Fairly common. Edible.

Distinguished from all other large *Lepiota* species by its reddening flesh, and additionally from *L. procera* in stockier appearance, differently shaped cap with coarse, shaggy, yellow-brown scales and lack of snake-like markings on the stem. It is also distinguished by its preference for growing in woodland rather than pastures.

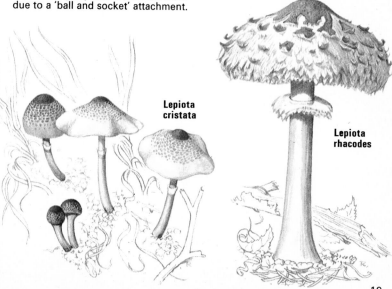

Lepiota
cristata

Lepiota
rhacodes

ARMILLARIELLA

Two species. Fruitbodies densely tufted at base of trunks. Parasitic.

ARMILLARIELLA MELLEA
Honey Fungus
Cap: 6-12 cm diam., convex, then flat to saucer-shaped, tan, tawny or cinnamon-brown, paler towards the indistinctly striate margin; ornamented with delicate, dark-brown, hair-like scales, prominent and crowded in young specimens, eventually disappearing except at centre in old fruitbodies.
Stem: 9-14 cm high, 1-2 cm wide, tough, cylindrical, pale-tawny, whitish at apex.
Ring: near apex of stem, thick, cottony, whitish, often with yellow flocci at margin.
Gills: dirty-whitish to flesh-coloured; adnate.
Flesh: white, soft. Taste acrid.
Spore-print: cream.
Habitat: at base of living and dead trunks or stumps. Autumnal. Very common. Edible when young.

CYSTODERMA

A genus of about 6 species. Fruitbody medium to small, caps yellow-ochre, orange-brown, reddish-brown or pale flesh-coloured with granular-mealy surface. Stems similarly coloured. Spore-print white.

CYSTODERMA AMIANTHINA
Syn: *Lepiota amianthina*
Cap: 2-3 cm diam., broadly bell-shaped with central boss, yellowish to ochre-brown, with granular mealy surface.
Stem: 4-6 cm high, 4-6 mm wide, same colour as cap, often darker at base, covered with mealy granules below ring, apex pale.
Ring: sometimes indistinct, same colour as cap, granular beneath.
Gills: adnate, cream.
Flesh: yellowish.
Spore-print: white.
Habitat: Heathy places or coniferous woodland, amongst grass and moss. Autumnal. Fairly common.

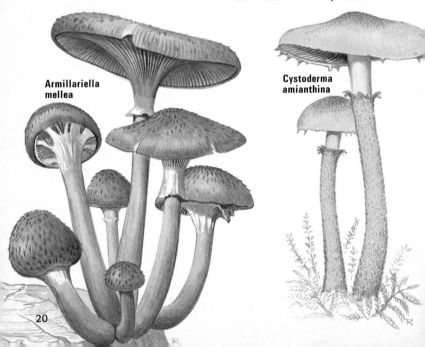

Armillariella mellea

Cystoderma amianthina

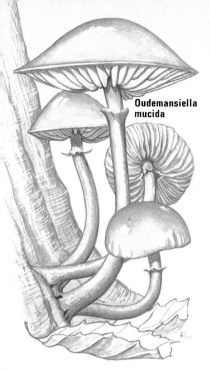

Oudemansiella
mucida

OUDEMANSIELLA

Four species. Cap viscid to glutinous but velvety in *O. longipes* and *O. badia*. Gills deep, distant.

OUDEMANSIELLA MUCIDA
Poached Egg Fungus
Syn: *Armillaria mucida; Collybia mucida*

Cap: 3-7 cm diam., soft, flabby, convex, very glutinous, white becoming flushed with grey, margin striate, almost translucent.

Stem: 5-7 cm high, 4-6 mm wide, often curved, tough, cartilaginous, cylindrical; often expanded disc-like at point of attachment; white to greyish.

Ring: spreading, white above, greyish below especially toward edge.

Gills: distant; deep, soft, white.

Spore-print: white.

Habitat: confined to beech, occurring in small clusters on trunks, branches, or on stumps. Autumnal. Common.

OUDEMANSIELLA RADICATA
Rooting Shank; Long Root
Syn: *Collybia radicata*

Cap: 5-8 cm., tough, convex, then flat, often coarsely radiately wrinkled around a central boss, sticky when wet, tacky when dry, varying in colour from pale buff or fawn to yellowish- or reddish-brown.

Stem: 10-18 cm high, 6-10 mm wide; tall, tough, cartilaginous, narrowly cylindrical, paler than cap, often greyish, longitudinally striate and usually twisted, prolonged below into a long root up to 30 cm in length.

Gills: distant, deep, white.

Spore-print: white.

Habitat: deciduous woodland, apparently terrestrial but arising from buried wood and roots etc. Late summer to autumn. Common.

Easily recognized by its sticky cap and tall elegant stem with long rooting base. *O. longipes* and *O. badia* both have a dry velvety brown cap and stem but otherwise resemble *O. radicata* in stature and habit.

Oudemansiella
radicata

**Craterellus
cornucopioides**

HYGROPHOROPSIS

Three species with several
varieties. Cap small to medium-
sized, funnel-shaped with white,
yellowish-ochre or orange suede-
like surface and incurved margin.
Gills crowded, similarly coloured,
regularly and repeatedly forked.
Spore-print white.

HYGROPHOROPSIS
AURANTIACA
False Chanterelle
Syn: *Cantharellus aurantiacus; Clitocybe aurantiaca.*
Cap: 3-6 cm diam., funnel-shaped
with enrolled margin, surface suede-
like, orange or yellowish-orange.
Stem: 2-4 cm high, 5-7 mm wide,
same colour as cap, often brownish
below when old.
Gills: decurrent, crowded, repeatedly forked, deep orange.
Smell: not distinctive.
Spore-print: white.
Habitat: coniferous woodland and
heaths. Autumnal. Common. Edible
but worthless.

CRATERELLUS

A single species. Cap tubular
with flared mouth, blackish.

CRATERELLUS
CORNUCOPIOIDES
Horn of Plenty; Trumpet of
Death
Cap: 2·5-7 cm high, tough, tubular,
horn-shaped with flaring mouth;
inner surface felty, sooty-brown to
blackish, drying out to pale grey-
brown; outer fertile surface smooth
to slightly uneven, dark slate-grey,
sometimes flushed ochre-coloured,
with ash-grey hoary appearance.
Habitat: amongst fallen leaves in
deciduous woodland, especially
beech. Autumnal. Occasional. Edible
and good.

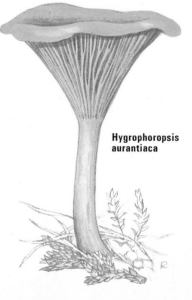

**Hygrophoropsis
aurantiaca**

Recognized by the orange, funnel-shaped fruitbody with strongly enrolled margin and the repeatedly forked crowded orange gills. Frequently mistaken from the genuine Chanterelle which is altogether more fleshy and. top-shaped, with irregularly branched fold-like shallow gills. It also has a smell of apricots.

CANTHARELLUS

Fruitbodies thin or fleshy, funnel-shaped with broad, shallow, irregularly branched gill-like folds. 'Gills' not sharp-edged like those of other mushrooms, but resembling veins or wrinkles. Spore-print white to cream.

CANTHARELLUS CIBARIUS
Chanterelle
Cap: 2·5-6 cm diam., top-shaped,

often slightly depressed at centre, gradually narrowed below into the short stalk; smooth, moist, entire fungus bright egg-yellow.

Stem: 2-6 cm high, 6-16 mm wide, short and squat.

Gills: decurrent, blunt, irregularly branched, fold-like and interconnected.

Smell: pleasant (of apricots).

Spore-print: white.

Habitat: deciduous woodland, especially on sandy or clay banks amongst moss. Autumnal. Occasional. Edible. Much sought after and collected for sale in markets. Easily dried for use in cooking.

The fleshy, egg-yellow fruitbody with irregularly branched, shallow, fold-like gills and smell of apricots is diagnostic. The only species liable to confusion with it is *Hygrophoropsis aurantiaca.* Here the fruitbody is much less fleshy, is funnel-shaped, orange and has true crowded gills which fork repeatedly in a regular manner.

Cantharellus cibarius

LACTARIUS
Milk Caps

A large genus of over 50 species. Cap small or robust, commonly funnel-shaped with dry or sticky, smooth or shaggy surface, variously coloured and frequently with conspicuous zones. Stem often narrowed below, sometimes pockmarked. Gills more or less decurrent, exuding a milky juice when cut or broken. In some species the milk is mild and tasteless in others exceedingly hot and peppery burning the tongue for some minutes after tasting – which it is quite safe to do. Smell often strong of curry powder or fenugreek in dried material.

LACTARIUS VELLEREUS

Cap: 8-16 cm diam., robust, funnel-shaped with enrolled margin, white, often patchily yellowish near edge, surface suede-like and rather shortly downy.

Stem: 4-6 cm high, 2-3 cm wide; short, squat, firm, whitish.

Gills: decurrent, distant, thick, white to pale-cream.

Milk: white, very hot and peppery.

Spore-print: white.

Habitat: Deciduous woodland. Autumnal. Occasional.

Readily recognized by large size, white cap with suede-like surface and white peppery milk.

L. piperatus differs in the very crowded gills and also in the less obvious and not at all downy suede-like texture of the cap surface.

Russula delica is a species which is also liable to confusion with this fungus but differs in having very brittle gills which fail to 'milk' when broken.

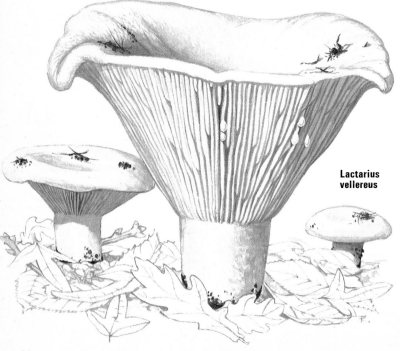

Lactarius vellereus

24

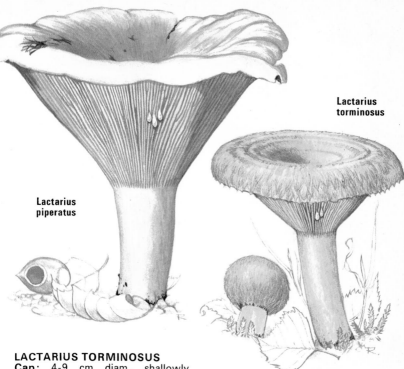

Lactarius torminosus

Lactarius piperatus

LACTARIUS TORMINOSUS
Cap: 4-9 cm diam., shallowly funnel-shaped, surface shaggy fibrillose, especially toward the enrolled margin, crushed strawberry pink with darker zones.
Stem: 5-6 cm high, 8-10 mm wide, pinkish flesh-coloured.
Gills: more or less decurrent, pale flesh-colour.
Milk: white, very peppery.
Spore-print: pale pinkish-buff.
Habitat: strictly with birch on heaths and commonland. Autumnal. Common.

The principal points for recognition of this species are the pink shaggy cap, the white peppery milk and its occurrence under birch.

A number of other species have shaggy caps and peppery milk. *L. pubescens* also grows with birch but in this particular species the cap is whitish with a thick shaggy covering devoid of zonation. *L. mairei*, a species with orange-buff colouring, is associated with oak.

LACTARIUS PIPERATUS
Cap: 10-16 cm diam., funnel-shaped, with smooth white surface.
Stem: 4-6 cm high, 2-3 cm wide; short, stocky, white.
Gills: white, decurrent, shallow, very densely crowded.
Milk: white, plentiful, very peppery.
Spore-print: white.
Habitat: deciduous woodland. Autumnal. Occasional.

This fungus can be fairly readily distinguished from *L. vellereus* by its smooth cap and densely crowded gills.

L. controversus is a similar species but has the cap blotched red or pink and has pale, rosy-buff gills. It is found associated with poplars or with dwarf willows on sand-dunes.

L. glaucescens is a species which is principally recognized by the milk slowly becoming faintly greenish-blue.

Lactarius
turpis

LACTARIUS TURPIS
The Ugly One
Syn. *L. plumbeus*
Cap: 8-20 cm diam., shallowly funnel-shaped or convex with depressed centre, surface sticky, smooth except for the enrolled felty margin, dark, dull olive-brown to almost black, brighter yellowish-olive towards edge.
Stem: 4·5-7 cm high, 2-2·5 cm wide; short, squat, sticky, same colour as, but paler than cap, often pitted.
Gills: more or less decurrent, dirty creamy-white becoming brown and discoloured when bruised.
Milk: white, plentiful, very peppery.
Spore-print: pale pinkish-buff.
Habitat: strictly associated with

birch on heaths and commonland, often overgrown with grass. Autumn. Very common.
The occurrence with birch, the large size, sombre colour, and white peppery milk afford easy recognition. An additional confirmatory character is that the cap and stem turn instantly deep violet when a drop of ammonia is applied; the gills react less strongly.

LACTARIUS GLYCIOSMUS
Coconut Smelling Milkcap
Cap: 2-6 cm diam., flat to slightly depressed, with a small central nipple, smooth, dry, pale greyish-lilac, sometimes dull-buff.
Stem: 4-5·5 cm high, 4-6 mm wide,

26

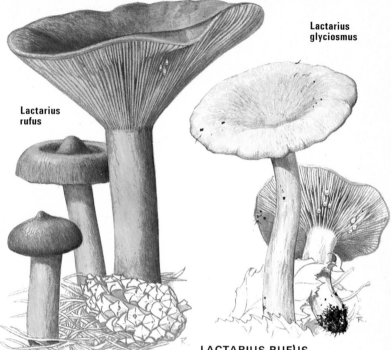

Lactarius glyciosmus

Lactarius rufus

same colour as cap but paler.
Gills: more or less decurrent, pinkish-buff to pale ochre.
Milk: white, mild becoming slightly peppery.
Smell: strong, sweetish, of desiccated coconut.
Spore-print: pale ochre-coloured.
Habitat: under birch on heaths and commonland. Autumnal. Fairly common.

The habitat, small greyish-lilac cap with central papilla and smell of coconut are the distinctive features of this fungus. *L. uvidus* has a tacky, violet-grey cap up to 10 cm diam., creamy to pinkish gills and a pale stem which becomes yellowish from below and finally marked with rusty spots. One striking feature of this species, however, is that the bitter milk, at first white, becomes violet as does the cut flesh. This fungus grows in damp birch woods.

LACTARIUS RUFUS

Cap: 4-6 cm diam., shallowly funnel-shaped with prominent central nipple, rich red-brown, surface smooth, dry, minutely grained, appearing falsely granular.
Stem: 4·5-5·5 cm high, 5-7 mm wide, same colour as cap but paler, base whitish.
Gills: decurrent creamy-flesh coloured.
Milk: white, plentiful, very peppery, but only after a minute or so.
Spore-print: pale pinkish-buff.
Habitat: coniferous woodland, very rarely with birch. Autumnal. Common.

Recognized from its habitat with conifers, red-brown cap with central nipple and very peppery milk. *L. hepaticus* is another species of coniferous woodland, but this has a smaller, liver-brown cap, 3-4 cm diam., light buff-coloured gills and an orange-brown stem. The milk is mild and dries yellow in a short time on a handkerchief.

LACTARIUS QUIETUS

Cap: 5-6·5 cm diam., shallowly convex with depressed centre, milky coffee colour with vague indication of darker zones, surface almost as if stippled.

Stem: 5-6 cm high, 6-10 mm wide, same colour as the cap but darker, especially toward base, and often redder.

Gills: decurrent, pale yellowish-flesh colour.

Smell: characteristic, oily.

Spore-print: pale pinkish-buff.

Habitat: strictly with oak. Autumnal. Common.

The association with oak, milky-coffee coloured cap, dark stem, mild tasting white milk and oily smell are the diagnostic characters.

LACTARIUS VIETUS

Cap: 4-7 cm diam., shallowly convex becoming depressed at centre, slightly sticky when moist, purplish-grey to lilac-flesh colour.

Stem: 5-7 cm high, 8-10 mm wide, same colour as cap but paler.

Gills: whitish to yellowish flesh colour, but brownish when bruised.

Milk: white, solidifying to pearl-grey spots on the gill, slowly peppery.

Spore-print: pale pinkish-buff.

Habitat: damp situations with birch. Autumn. Occasional.

The purplish-grey cap, the white milk solidifying to pearl-grey spots and occurrence in damp birchwoods are the important features for recognition. *L. glyciosmus* occurs in similar situations but is usually smaller, the milk does not solidify as pearl-grey spots and it has a characteristic smell of desiccated coconut.

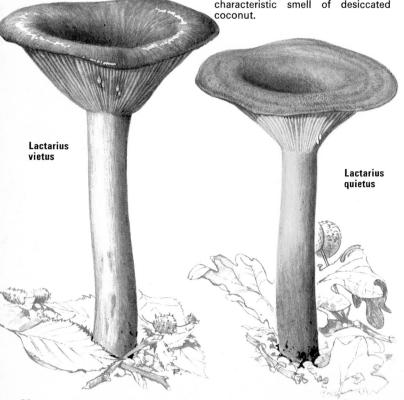

Lactarius
vietus

Lactarius
quietus

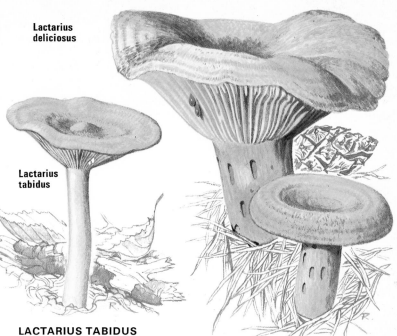

Lactarius deliciosus

Lactarius tabidus

LACTARIUS TABIDUS
Cap: 2·5-4 cm diam., flat or slightly depressed, irregularly radially wrinkled about a small central nipple, orange-brown paling to yellowish-buff when dry, margin somewhat striate when moist.
Stem: 2-4 cm high, 4-6 mm wide, same colour as cap.
Gills: decurrent, buff.
Milk: mild, white, changing to yellow if allowed to dry on a handkerchief.
Spore-print: pale buff.
Habitat: deciduous woodland. Autumnal. Common.

The small orange-brown cap and mild milk drying yellow on a handkerchief are the main diagnostic characters of *L. tabidus.* There are several similar species with yellowing milk. *L. hepaticus,* a common pinewood species has a liver-coloured cap; *L. britannicus* is a slightly larger fungus of beechwoods with cap ranging from dark bay-brown around the papillate centre, through bright tawny-chestnut to creamy-orange at the margin.

LACTARIUS DELICIOSUS
Cap: 6-10 cm diam., convex with depressed centre or shallowly funnel-shaped, moist, reddish-orange with darker greenish zones and variable development of green staining, surface appearing almost as if granular-stippled.
Stem: 6-8 cm high, 1·5-2·0 cm wide, orange, often pitted.
Gills: orange-yellow, staining greenish.
Milk: brilliant carrot-colour eventually becoming wine-red in 30 minutes.
Spore-print: pale pinkish-buff.
Habitat: coniferous woodland, especially pine. Autumnal. Common. Edible.

An unmistakable species due to its orange colour and green staining of all parts, and the presence of vivid carrot-coloured milk.

Other species of *Lactarius* with brightly coloured milk include *L. chrysorrheus,* a species associated with oak in which the milk becomes quickly yellow on the gill.

29

CLITOCYBE

About 60 species ranging from large and robust to medium-sized fungi, mostly dull coloured and either top-shaped or funnel-shaped, with somewhat decurrent to strongly decurrent gills. Spore-print white, cream or even with pinkish flush. Species of *Omphalina* are very similar, with decurrent gills, but are smaller.

CLITOCYBE NEBULARIS
The Clouded Clitocybe
Cap: 7-16 cm diam., robust, fleshy, convex with a low central hump to more or less flat, cloudy-grey sometimes with a brown tinge, surface dry appearing to have a faint hoary bloom.
Stem: 7-10 cm high, 1·5-2·5 cm wide, cylindrical, same colour as cap but paler, often striate.
Gills: decurrent, crowded, dirty creamy-white.
Smell: characteristic, unpleasant.
Spore-print: creamy-white.
Habitat: deciduous woodland, especially in areas rich in humus, near piles of rotting leaves or grass, usually gregarious and sometimes forming fairy-rings. Autumnal. Common. POISONOUS.

Recognized by its cloudy-grey colour and characteristic smell.

CLITOCYBE ODORA
The Sea Green Clitocybe
Cap: 3-5 cm diam., flat to slightly depressed at centre, entire fungus beautifully blue-green.
Stem: 3·5-4·5 cm high, 6-8 mm wide.
Gills: slightly decurrent, paler than cap.
Smell: strong, sweet, fragrant, of aniseed.
Spore-print: white.
Habitat: deciduous woodland, often gregarious in small groups. Autumnal. Occasional.

Readily identified by its sea-green colour and strong fragrant smell.

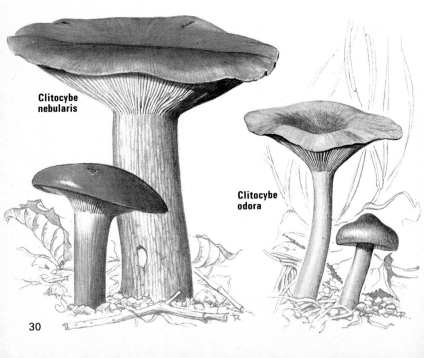

Clitocybe nebularis

Clitocybe odora

Clitocybe
vibecina

CLITOCYBE VIBECINA
Cap: 2·5-4 cm diam., convex with depressed centre to funnel-shaped, when moist uniformly watery grey-brown with striate margin, but drying to a pale opaque biscuit colour with a dark brown spot in the base of the funnel.
Stem: 3-4 cm high, 5-6 mm wide, greyish with white woolly base.
Gills: decurrent, pale grey-brown.
Smell: of meal when crushed.
Spore-print: white.
Habitat: deciduous and coniferous woodland, also heathland under bracken. Autumnal but persisting well into winter. Common.

There are several closely-related, small, grey-brown species of which *C. vibecina* is the most frequent. An important feature of note is the completely different appearance of the moist and dry states of these fungi.

C. dicolor differs in lack of mealy smell and in the stem being proportionally longer and distinctly darker at the base. *C. phyllophila* has a lead-white cap up to 7 cm diam., which soon discolours, becoming creamy-coloured.

CLITOCYBE FLACCIDA
Syn : *Clitocybe inversa*
Cap: 4-7 cm diam., tough, elastic, funnel-shaped with enrolled margin, surface smooth, tan, becoming somewhat reddish brown when old but paling when dry.
Stem: 4-6 cm high, 5-7 mm wide, tough, same colour as cap, base woolly.
Gills: decurrent, crowded, pale to yellowish.
Spore-print: white to pale cream.
Habitat: deciduous woodland, often gregarious or clustered sometimes in fairy-rings. Autumnal but persisting well into winter. Fairly common.

A similar fungus of coniferous woods is sometimes distinguished as *C. gilva* (Syn: *C. splendens*) and has an altogether paler yellowish cap. Both species have tiny spherical roughened spores.

C. infundibuliformis is another tough-elastic species of deciduous woodland. Flesh pale to pale tan-coloured, funnel-shaped cap, 4-5 cm diam., a firm, rather narrow stem, similarly coloured, about 5 mm wide and whitish decurrent gills.

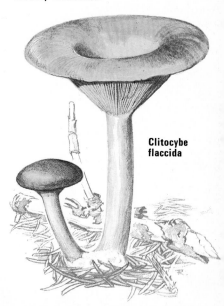

Clitocybe
flaccida

RUSSULA
Brittle Gills

Probably the largest genus with over 120 species. Difficult to identify, it is essential to taste each specimen to note whether it is hot or mild, and also to take a spore-print on glass to match the exact shade somewhere between pure white and ochre. The species, often brilliantly coloured in shades of pink, red, purple, orange, yellow, green etc, range from small to robust and have caps which are mostly shallowly convex with shortish fragile stems. The easiest generic character to recognize is the brittle nature of the gills when the finger is lightly run over them. Further, with few exceptions (*R. nigricans* and relatives), the gills are all of the same length, practically without intervening short gills.

RUSSULA NIGRICANS

Cap: 9-15 cm diam., convex then depressed, pale then rapidly patchily brown becoming uniformly dark-brown, finally almost black.
Stem: 5-7 cm high, 2-3 cm wide, becoming brown.
Gills: thick, adnate, exceptionally distant, extremely brittle with intervening short gills, dirty creamy-white, red where bruised.
Flesh: firm, white but reddening and finally blackening when broken.
Habitat: deciduous woodland, especially beech. Autumnal. Common.

The sombre colour and exceptionally distant gills with intervening short gills make this an easy species to recognize. It often persists in a completely black rotten condition for many months and may be parasitized by one or other of two small whitish or greyish agarics: *Nyctalis parasitica* or *Nvctalis asterophora* (see p. 54). *Russula acrifolia* and *R. adusta* are similar, but have crowded, sub-decurrent gills and intervening short gills.

Russula
nigricans

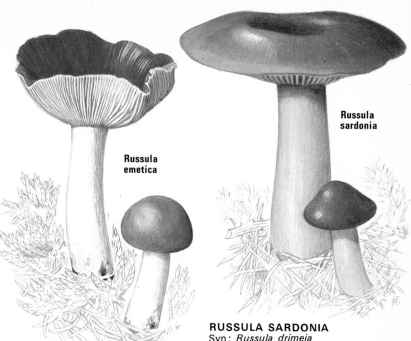

Russula emetica

Russula sardonia

RUSSULA EMETICA
The Sickener
Cap: 4-7 cm diam., convex then depressed, brilliant scarlet with a moist shiny surface, margin eventually coarsely striate.
Stem: 5-8 cm high, 1·5-2 cm wide, rather tall, fragile, pure white, the lower portion usually somewhat club-shaped.
Gills: adnexed, white.
Flesh: very hot.
Spore-print: pure white.
Habitat: coniferous woodland. Autumnal. Common. May cause sickness if eaten raw.

The habitat, brilliant colour, white spore-print, rather tall stature and hot peppery taste are the vital characters for correct identification. There are a number of closely related forms with slightly differing micro-characters. *R. mairei* is almost indistinguishable, except in growing with beech and having a shorter, firmer cylindrical stem.

RUSSULA SARDONIA
Syn: *Russula drimeia*
Cap: 6-10 cm diam., convex to broadly bell-shaped, varying from dark reddish- or violet-purple to purplish-black.
Stem: 7-12 cm high, 1·5-2 cm wide, rather tall, beautifully purple.
Gills: adnexed, primrose, often with watery droplets along the edge.
Flesh: white, very hot.
Spore-print: cream.
Habitat: coniferous woodland. Autumnal. Common.

R. queletii, another pinewood species, although much rarer, is virtually indistinguishable except for the whitish to very pale-cream gills. *R. caerulea* also occurs with pines and has a similarly coloured cap with shiny surface, although with a prominent central nipple, but the flesh is mild or only slightly peppery, the stem is white and the spore-print ochre. *R. atropurpurea* has a dark purplish-red to almost black cap but occurs in deciduous woodland. It has whitish gills, a somewhat peppery taste, and a white to off-white spore-print.

RUSSULA ATROPURPUREA

Cap: 4-10 cm diam., convex sometimes with slight hump, moist, dark reddish-purple to almost black, old specimens often mottled yellow at centre.
Stem: 5-7 cm high, 1·5-2 cm wide, short, squat, white, with rust-coloured base.
Gills: adnexed, whitish to palish-cream, often discoloured with rusty spots.
Flesh: mild to slightly peppery.
Spore-print: white to off-white.
Habitat: deciduous woodland. Autumnal. Common.

Amongst species of deciduous woods, *R. atropurpurea* is distinguished by its almost black cap and off-white spore-print. *R. fragilis* is similar but altogether smaller and more delicate with very peppery flesh and a white spore-print. The cap, seldom more than 4 cm diam., is patchily purplish pink at the margin with a dark depressed centre.

Dark purplish forms of *R. xerampelina* can be separated by having a smell of crab, a pale ochre spore-print and, a green reaction on the stem when rubbed with a crystal of Ferric Alum.

RUSSULA SANGUINEA

Cap: 7-10 cm diam., shallowly convex, then flat often with slight hump at centre, blood red, with a moist, almost granular look.
Stem: 8-12 cm high, 1·5-2 cm wide, whitish, flushed strongly pink.
Gills: slightly decurrent, crowded, ivory.
Flesh: white, hot.
Spore-print: cream.
Habitat: coniferous woodland, autumnal. Occasional.

The colour, strong pink flush on the stem, hot taste, cream spore-print and association with pines are the important points for recognition.

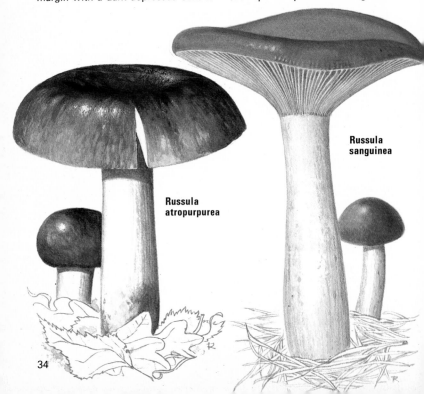

Russula atropurpurea

Russula sanguinea

34

RUSSULA OCHROLEUCA

Cap: 5-8 cm diam., shallowly funnel-shaped, moist, bright ochre-yellow to greenish-yellow.
Stem: 6-8 cm high, 1·5-2 cm wide, rather soft with firmer rind, white, eventually greyish, surface ornamented with faint, densely-crowded short, raised longitudinal lines.
Gills: whitish.
Flesh: mild to moderately hot.
Spore-print: pale-cream.
Habitat: coniferous and deciduous woodland. Autumnal but persisting late in the season. Very common.

R. ochroleuca is one of the commonest Russulas and fairly easy to recognize although liable to confusion with R. fellea. The latter is quite frequent in beech woods and is usually smaller, more compact and more brownish in colour. The cap has a smooth, oiled or waxed appearance and ranges from pale straw at the margin to deep straw at the centre. The gills and stems are similarly coloured but paler and the fungus has a fruity smell resembling apples.

RUSSULA CAERULEA

Cap: 4-7 cm diam., convex with central nipple, shiny, uniformly dark purplish-violet.
Stem: 6-8 cm high, 1-1·5 cm wide, tall, white.
Gills: cream then yellow.
Flesh: mild.
Spore-print: ochre.
Habitat: coniferous woodland. Autumnal. Occasional.

The very dark cap with central nipple, together with yellow gills and ochre spore-print, white stem and coniferous habitat are the specific characters to note. Russula sardonia and R. queletii, both species of coniferous woods, have similarly coloured caps but they lack a central nipple, the gills and spore-print are much paler and the stems are pink. The taste is hot or very hot. Dark-purplish forms of Russula xerampelina can be separated by their smell of crab.

Russula
caerulea

Russula
ochroleuca

35

RUSSULA FELLEA

Russula fellea

Cap: 4-5 cm diam., convex, varying from pale straw at the margin to darker straw at centre, surface with oiled or waxed appearance.

Stem: 4-6 cm high, 1·5-1·8 cm wide, same colour as cap but paler.

Gills: adnexed, same colour as cap.

Flesh: peppery, with a smell of apples or geraniums.

Spore-print: pale-cream.

Habitat: deciduous woodland, especially beech. Autumnal. Fairly common.

This species is similar to *R. ochroleuca* but is distinguished by the more brownish colour of the entire fruitbody and the smell of apples, as well as being more compact.

R. claroflava is a species associated with birch, often growing in sphagnum moss around margins of lakes and in swampy situations. It has a chrome-yellow cap, cream gills and the entire fungus including the flesh slowly blackens.

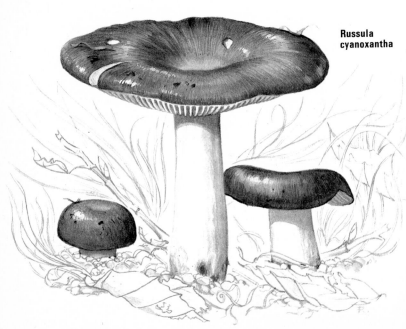

Russula cyanoxantha

RUSSULA FOETENS

Cap: 9-14 cm diam., almost globular in the young stage, but later convex, slimy to glutinous, dingy ochre-brown, with conspicuously grooved margin.
Stem: 10-12 cm high, 2-4 cm wide, hollow, rather brittle, much paler than cap.
Gills: often weeping with watery droplets along edge, dingy white flushed with pale straw colour.
Flesh: peppery, with strong, foetid, oily smell.
Spore-print: cream.
Habitat: deciduous woodland. Autumnal. Fairly common.

The robust habit, dull brown cap with grooved margin, and the strong smell are diagnostic. *R. laurocerasi* differs in having a less strong smell of bitter almonds or crushed leaves of Cherry Laurel. *Russula sororia*, found under oaks, is a smaller, flatter species with greyish-brown cap and grooved, tuberculate margin.

RUSSULA CYANOXANTHA

Cap: 7-10 cm diam., convex, slightly depressed at centre, moist, dark greyish-purple with olive tones.
Stem: 8-11 cm high, 1·5-2 cm wide, white, hard.
Gills: white, softly pliable.
Flesh: mild.
Spore-print: white.
Habitat: deciduous woodland. Autumnal. Common.

The purplish-grey cap with olive tones and elastic white gills are sure aids to identification. The gills are much less brittle than those of other *Russula* species.

Russula parazurea and *R. ionochlora* are similar but smaller species with a distinct hoary bloom to the cap and a cream spore-print. *R. heterophylla* and *R. aeruginea*, both associated with birch, have green caps; the former is rather rare and has a white print, the latter common with a buff-coloured print.

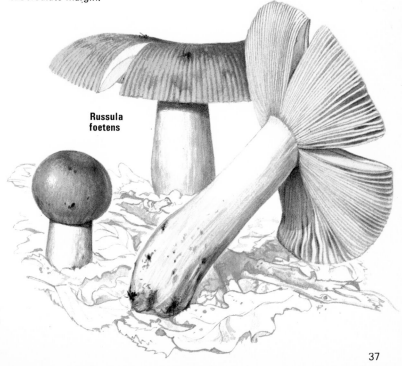

Russula foetens

FLAMMULINA

A single species characterized by
lignicolous habit, bright tan-
coloured, sticky cap, dark-brown
velvety stem and white spore-
print.

FLAMMULINA VELUTIPES
The Velvet Shank
Cap: 2·5-5 cm diam., shallowly
convex becoming flat, bright-yellowish or orangey-tan, often darker and
more brownish at the centre, moist
becoming sticky when wet, shiny
when dry.
Stem: 2·5-5 cm high, 4-6 mm
wide, very dark-brown and conspicuously velvety, paling to yellow
nearer the cap.
Gills: adnexed, rather distant, pale
creamy-yellow.
Habitat: on trunks and branches,
especially of dead elm, often in
considerable quantity, forming small
tiered clusters, very rarely growing
from buried roots. Late autumn and
winter.

Flammulina
velutipes

TRICHOLOMOPSIS

Three lignicolous species.
Fruitbodies tough or fleshy, often
rather robust with distinctly or
indistinctly felty surface, which
may be brightly coloured, yellow,
purple etc. Spore-print white.

TRICHOLOMOPSIS RUTILANS
Plums and Custard
Syn: *Tricholoma rutilans*
Cap: 6-12 cm diam., convex to
broadly bell-shaped, yellow, densely
covered with tiny fleck-like purple
scales which are continuous at
centre but nearer the margin become
pulled further apart to show more and
more of the yellow background.
Stem: 6-9 cm high, 1-1·5 cm wide,
similar in colour to cap, likewise
densely flecked below with purple
scales which disappear towards the
yellow apex.
Gills: yellow.
Flesh: yellowish.
Spore-print: white.
Habitat: on conifer stumps. Autumnal. Fairly common.
The combination of yellow and
purple colours, and occurrence on
conifer stumps makes this fungus
easy to recognize.
T. platyphylla is a tough grey-
brown fungus with large flabby cap
5-12 cm diam., which is rather
streaky and slightly fibrillose; the gills
are very deep, distant and whitish.

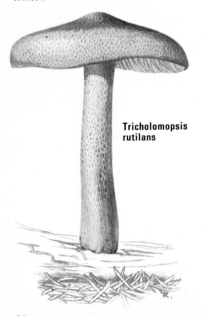

Tricholomopsis
rutilans

LEPISTA

Six species. Fruitbodies mostly fairly robust and fleshy and in some species bright lilac in parts. Gills more or less sinuate. Spore-print pale pinkish.

LEPISTA SAEVA
Blewit
Syn : *Tricholoma saevum; T. personatum; L. personata*
Cap: 6-8 cm diam., convex then flat, moist, varying from buff to greyish-buff.
Stem: 5-6 cm high, 1·5-2 cm wide, often enlarged at base, bright violet with streaky fibrillose surface.
Gills: whitish to pale flesh-coloured.
Spore-print: pale pinkish.
Habitat: in grassland, often forming fairy-rings. Autumnal. Uncommon. Edible and good, sometimes sold in shops.

Similar to *L. nuda* which differs in its woodland habitat and in the bright violet colour of all parts of the fruitbody.

LEPISTA NUDA
Wood Blewit
Syn : *Tricholoma nudum*
Cap: 6-10 cm diam., shallowly convex, becoming flat, often with greasy or water-soaked appearance, varying in colour from entirely violet to reddish-brown with violet tint localized to margin.
Stem: 6-8 cm high, 1·5-2 cm wide, bright bluish-lilac.
Gills: at first vivid violet becoming pinkish with age.
Spore-print: pale pinkish.
Habitat: deciduous woodland, compost heaps. Late autumn persisting into winter. Common. Edible and good, sometimes sold in shops.

Unlikely to be confused with other mushrooms provided a check is made on the colour of the spore-print. There are many *Cortinarius* species which show varying amounts of lilac or violet tints but these all have a rusty-brown print.

Lepista saeva

Lepista nuda

LACCARIA

Seven to eight species. Cap small to medium, often brightly coloured and usually with scurfy or felty surface which assumes a completely different aspect when dry to that when wet. Gills rather thick, distant, waxy and pinkish or deep amethyst. Spore-print white.

Laccaria amethystea

LACCARIA AMETHYSTEA
Amethyst Deceiver
Syn. *L. amethystina*
Cap: 2·5-4 cm diam., convex then shallowly convex, often depressed at centre, surface scurfy-felty, when moist entire fungus bright violet, but when dry this fades to pale buff with faint lilac tint.
Stem: 4-6 cm high, 6 mm wide, deep violet, but paler when dry.
Gills: deep, thick, distant, adnate or adnexed, bright violet fading to lilaceous-flesh-colour when dry.
Spore-print: white.
Habitat: deciduous woodland. Autumnal. Common.

Easily recognized by its habit and the brilliant violet colour of the entire fruitbody when wet, less readily identified when dry.

Laccaria laccata

LACCARIA LACCATA
The Deceiver
Cap: 2-4 cm, shallowly convex to broadly bell-shaped, sometimes slightly depressed at centre; surface felty or scurfy; when moist bright red-brown with striate margin, drying out to pale buff and opaque and not striate.
Stem: 4-5 cm high, 5 mm wide, tough, streaky fibrillose, reddish-brown, often twisted.
Gills: deep, thick, distant, waxy, pinkish flesh-colour with waxy appearance.
Spore-print: white.
Habitat: deciduous woodland and coniferous woodland, heaths. Often gregarious. Autumnal. Very common.

Because of its variability in appearance due to the degree of moistness of the cap, it can be very difficult to recognize in all its guises, hence the common name. The thick, distant, waxy, flesh-coloured adnate or adnexed gills are a good guide to to recognition. *L. proxima* is a more robust species of damp boggy situations with stem up to 8 cm high.

40

COLLYBIA

Approximately 36 species, but the small dingy, grey–brown species with greyish gills are now often separated into the genus *Tephrocybe*.

COLLYBIA MACULATA
Foxy Spot
Cap: 5-9 cm diam., convex, white with pinhead-sized or larger red-brown spots, often becoming entirely pale red-brown with age.
Stem: 8-10 cm high, 1-1·6 cm wide, firm, white, often longitudinally striate, tapering below into a short rooting base.
Gills: densely crowded, shallow, pale-cream spotted with red-brown.
Habitat: coniferous woodland or amongst bracken in heathy situations where it is often gregarious and sometimes forms fairy-rings. Autumnal. Common.

The occurrence in coniferous woods or on heathland of the gregarious, tough white fruitbodies with foxy-red spotting of all parts and very crowded gills makes for easy identification.

COLLYBIA FUSIPES
Spindle Shank
Cap: 3-7 cm diam., broadly bell-shaped with central boss, smooth, dark red-brown drying pinkish buff or pale tan.
Stem: 8-10 cm high, 1-1·5 cm wide, tough, similarly coloured to cap, gradually enlarged below then conspicuously narrowed into a rooting portion, surface distinctly grooved.
Gills: adnexed, broad, distant, whitish then flushed with red-brown, also often spotted with brown.
Spore-print: white.
Habitat: tufted at base of tree trunks, especially oak. Late summer to early autumn. Occasional.

The rooting bases of the fruitbodies fuse together below ground and can be traced back to a black woody branched structure. The dark liver-coloured cap and tough grooved rooting stem are the distinctive characters.

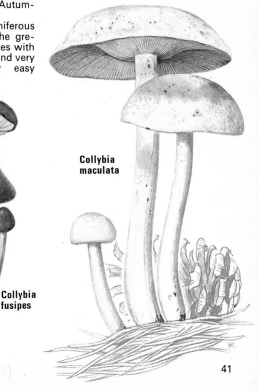

Collybia maculata

Collybia fusipes

**Collybia
peronata**

COLLYBIA DRYOPHILA
Syn : *Marasmius dryophilus*
Cap : 2-3 cm diam., shallowly convex to flat, smooth, pale biscuit-colour, deeper brown at centre, drying out to almost white with hint of yellowish brown at centre only.
Stem : 4-6 cm high, 2-4 mm wide, tall, narrow, tough, smooth, varying from yellowish- to orange-brown.
Gills : adnexed, crowded, whitish.
Spore-print : white.
Habitat : deciduous woodland or in grassy rides. May to November. Very common.

COLLYBIA CONFLUENS
Syn : *Marasmius confluens*
Cap : 2-4 cm diam., shallowly bell-shaped to flat, very thin, leathery, surface hoary, pale biscuit colour sometimes with hint of flesh-colour, almost white when dry.
Stem : 6-7 cm high, 3 mm wide, tall, thin, often flattened, very tough, entirely covered with minute whitish hairs, otherwise same colour as cap.
Gills : adnexed, very densely crowded, shallow, tinged colour of cap.
Spore-print : white.
Habitat : forming dense tufts of up to 10 or 15 fruitbodies in deep leaf litter in deciduous woodland, especially beech. Autumnal. Occasional.

COLLYBIA PERONATA
Wood Woolly Foot
Syn : *Marasmius peronatus*
Cap : 4-6 cm diam., very broadly bell-shaped to flat, sometimes with central boss, smooth, very tough and leathery, ochre-coloured to reddish-brown, drying paler.
Stem : 7-9 cm high, 5 mm wide, tall, narrow, but very tough, pale yellowish-buff, thickly covered toward base with pale yellowish woolliness.
Gills : tough, leathery, distant, adnexed, separating from around the top of the stem in a false collar; same colour as cap.
Flesh : thin, leathery, yellowish, taste peppery.
Spore-print : white.
Habitat : especially deciduous woodland amongst leaf litter. Autumnal. Common.
   The very thin leathery texture of the fruitbody makes it impossible to break; it has to be torn apart.

**Collybia
dryophila**

MARASMIUS

About 30 species, mostly small to medium-sized and tough, with a white spore-print.

MARASMIUS OREADES
Fairy-Ring Champignon
Cap: 3-5 cm diam., convex with broad central boss, tough, smooth, pinkish-tan drying to pale buff, margin often grooved.
Stem: 4·5-5·5 cm high, 2·5-3 mm wide, firm, tough, pale buff.
Gills: adnexed, deep, distant, whitish.
Spore-print: white.
Habitat: pastures, lawns, roadside verges and the commonest cause of 'fairy-rings' in turf. Late summer to autumn. Common. Edible.

Recognized when forming fairy-rings by the tough, broadly bell-shaped, buff-coloured cap with distant gills. However it does not always grow in rings.

Collybia confluens

Marasmius oreades

43

TRICHOLOMA

About 55 terrestrial species, mostly rather fleshy with sinuate gills and a white spore-print. The caps may be sticky or dry, smooth, felty or scaly and can vary in colour from bright yellow, yellowish-green, pink, violet or brown, to grey. Gills are likewise variable in colour, making it essential to check the colour of a spore-print. The stem may be smooth, fibrillose or scaly. In *T. cingulatum*, together with a very few other rare species, there is a distinct ring.

TRICHOLOMA GAMBOSUM
St George's Mushroom
Cap: 5-10 cm diam., shallowly convex, often with undulating margin, smooth, white to very pale buff at centre.
Stem: 4-7 cm high, 1·5-2 cm wide, short, squat, white.
Gills: sinuate, crowded, white.
Flesh: thick, smelling strongly of meal.
Spore-print: white.
Habitat: in pastures, roadside verges, hedge-bottoms. Appears in spring, usually around St George's Day (April 23rd), hence the common name. Edible.

This whitish mushroom is readily recognized by its squat, fleshy fruit-bodies which occur in spring. *T. album* and *T. lascivum* are whitish autumnal species.

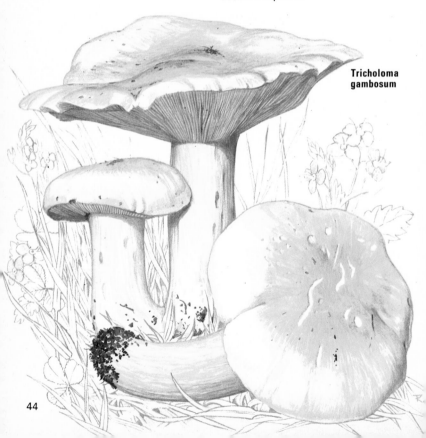

Tricholoma gambosum

TRICHOLOMA SULPHUREUM
Gas Tar Fungus

Cap: 4-6 cm diam., convex to broadly bell-shaped, smooth, sulphur-yellow.

Stem: 7-8 cm high, 1-1·5 cm wide, same colour as cap.

Gills: sinuate, deep, distant, rather thick, same colour as cap.

Flesh: yellow with pungent smell of gas tar.

Spore-print: white.

Habitat: deciduous woodland. Autumnal. Occasional.

Easily identified by the smell and uniform sulphur-yellow colouring of the entire fruitbody.

Tricholoma
terreum

Tricholoma
sulphureum

nipple, mouse-grey with a felty surface.

Stem: 4-7 cm high, 7-10 cm wide, whitish.

Gills: sinuate, distant, whitish or flushed grey.

Flesh: whitish or watery-grey, mild tasting, without distinctive smell.

Spore-print: white.

Habitat: woodland, especially with conifers. Autumnal. Fairly common. Edible.

The felty grey cap and pale, distant, sinuate gills are distinctive characters.

T. argyraceum is superficially similar but smells of meal when crushed and in old specimens the gills turn yellow at the onset of decay. *T. cingulatum*, also similar and with mealy smell, is instantly recognized by having a cottony ring on the stem — one of the very few members of the genus to have one.

TRICHOLOMA TERREUM
Cap: 3-6 cm, broadly bell-shaped, sometimes slightly depressed at centre, with a rather prominent

HYGROPHORUS

A large genus of about 90 species, many of which occur in grassland and are often of vivid colour – red, orange, yellow, pink, green, and have either a dry or glutinous surface. The cap is either shallowly convex, broadly bell-shaped, top-shaped or acutely conical. In the latter instance it often bruises black on handling or in age. Gills are strongly decurrent in many species but in others narrowly adnexed or free, but they have a waxy look. The majority of species are found in open situations, growing in short turf, but some grow in woodland.

HYGROPHORUS CONICUS
Cap: up to 3 cm high, acutely conical with fibrillose surface, yellow or orange becoming black on handling or with age.
Stem: up to 6 cm high, 5 mm wide, fibrillose, yellow, then blackening.
Gills: ascending, almost free, pale yellow.
Spore-print: white.
Habitat: in grassland. Autumnal. Fairly common.

The acutely conical fibrillose orange cap which, like the rest of the fruit-body blackens when handled, is very distinctive.

The only species which is liable to confusion with this species is *H. nigrescens.* This has a predominantly scarlet-red cap and is more robust than *H. conicus. H. conicoides* occurs on sand dunes and has a cherry-red cap.

Hygrophorus
conicus

46

HYGROPHORUS PUNICEUS

Cap: 5-10 cm diam., broadly bell-shaped, smooth, scarlet but soon fading and becoming more yellowish with age.

Stem: 7-10 cm high, 1-1·5 cm wide, fibrillose, same colour as cap, but white at base.

Gills: adnexed, yellow flushed red.

Spore-print: white.

Habitat: grassland. Autumnal. Occasional.

Differs from *H. coccineus* in having a more robust appearance. *H. splendidissimus* is a similar species, but is said to have a smooth stem and also differs in the colour of the flesh in the centre of the stem: it is yellow (not whitish).

HYGROPHORUS COCCINEUS

Cap: 2·5-5 cm diam., broadly bell-shaped, sometimes flat or even concave with central boss, smooth, bright scarlet fading with age to yellowish.

Stem: 2·5-6 cm high, 5-10 mm wide, same colour as cap but yellow at base.

Gills: adnate with decurrent tooth, rather distant, red.

Spore-print: white.

Habitat: in grassland. Autumnal. Occasional.

H. puniceus is a much more robust species with cap up to 10 cm diam., and has a white base to the stem. *H. miniatus* has a red cap with minutely scurfy surface.

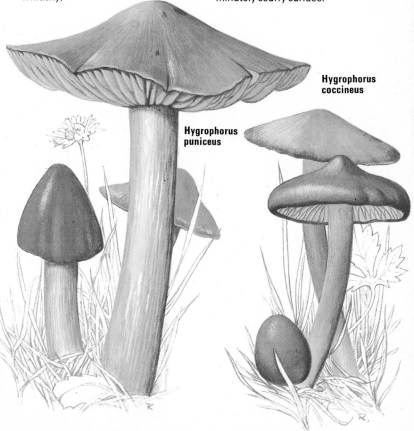

Hygrophorus
coccineus

Hygrophorus
puniceus

47

HYGROPHORUS PRATENSIS

Cap: 4-6 cm diam., broadly bell-shaped to top-shaped with a central boss, smooth, pale-tan to buff.
Stem: 5-6 cm high, 6-8 mm wide, same colour as cap.
Gills: decurrent, distant, pale buff.
Spore-print: white.
Habitat: in grassland. Autumnal. Occasional. Edible.

Distinguished by buff colours and distant decurrent gills.

HYGROPHORUS NIVEUS

Cap: 1·5-2·5 cm diam., broadly bell-shaped to top-shaped, white and watery with striate margin when moist, opaque when dry.
Stem: 2-3 cm high, 4-5 mm wide, tapering downward; white.
Gills: decurrent, distant; white.
Habitat: in grassland. Autumnal. Common.

H. virgineus is similar but usually more robust. *H. russo-coriaceus* has a more creamy tint and a strong smell of Russian leather or of pencil sharpenings.

MYCENA

A large genus of over 100 species, mostly small with delicate conical to bell-shaped caps borne on a fragile elongated stem. Some species are more robust, with caps up to 4 cm in diameter and rather leathery (*M. galericulata*). A few species exude a white, red or orange latex when the stem is broken. Gills uniformly coloured or with a dark dotted gill-edge.

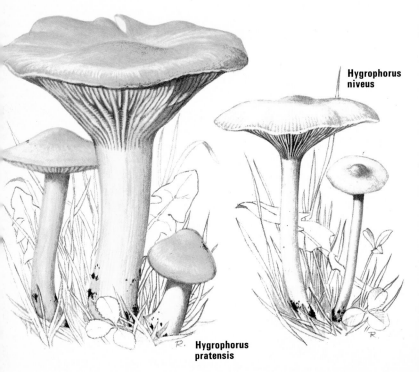

Hygrophorus niveus

Hygrophorus pratensis

Mycena galericulata

MYCENA GALERICULATA
The Leathery Mycena
Cap: 2-4·5 cm diam., broadly bell-shaped, flat with a central boss, tough, leathery, varying from grey-brown to buff, margin striate to somewhat grooved.

Stem: 7-10 cm high, 3-5 mm wide, tough, cartilaginous, smooth, polished, grey-brown.

Gills: adnate with decurrent tooth, deep, distant, interveined, whitish then flesh-coloured.

Smell: mealy when crushed.

Spore-print: white.

Habitat: clustered on stumps, or from buried wood. Typically autumnal, but occurring sporadically throughout the year. Very common.

Mycena inclinata differs in cap colour and tawny base to stem. *M. polygramma* has a dark smoky-brown cap which is broadly bell-shaped, 1·5 cm diam., with a central nipple and grooved margin, and firm elongated narrow stem up to 10 cm high which is silvery blue-grey and conspicuously striate. It grows singly on stumps and twigs.

MYCENA INCLINATA
Cap: 1·5-2·5 cm diam., broadly bell-shaped, fairly tough, dull date-brown or grey-brown, striate at margin which over-reaches the gills and is slightly toothed.

Stem: up to 7 cm high, 3 mm wide, at first silvery blue-grey and striate but fading, becoming dark tawny brown from base upward.

Gills: ascending, adnate, whitish, then flesh-colour.

Smell: distinctive, rancid.

Spore-print: white.

Habitat: densely tufted on old stumps, almost exclusively on oak, but reported on sweet chestnut and there is one authentic record on plum.

The tufted habit on oak stumps, the small, dull brown caps borne on stems with a tawny-brown base are reliable features for identification.

Mycena inclinata

49

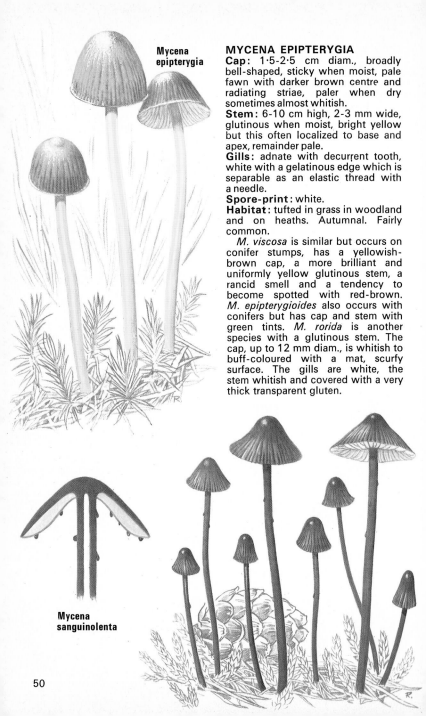

Mycena epipterygia

MYCENA EPIPTERYGIA

Cap: 1·5-2·5 cm diam., broadly bell-shaped, sticky when moist, pale fawn with darker brown centre and radiating striae, paler when dry sometimes almost whitish.

Stem: 6-10 cm high, 2-3 mm wide, glutinous when moist, bright yellow but this often localized to base and apex, remainder pale.

Gills: adnate with decurrent tooth, white with a gelatinous edge which is separable as an elastic thread with a needle.

Spore-print: white.

Habitat: tufted in grass in woodland and on heaths. Autumnal. Fairly common.

M. viscosa is similar but occurs on conifer stumps, has a yellowish-brown cap, a more brilliant and uniformly yellow glutinous stem, a rancid smell and a tendency to become spotted with red-brown. *M. epipterygioides* also occurs with conifers but has cap and stem with green tints. *M. rorida* is another species with a glutinous stem. The cap, up to 12 mm diam., is whitish to buff-coloured with a mat, scurfy surface. The gills are white, the stem whitish and covered with a very thick transparent gluten.

Mycena sanguinolenta

MYCENA SANGUINOLENTA

Cap: 7-10 mm high, narrowly conical, dark red-brown with striate margin.

Stem: 3-5 cm high, 1-1·5 mm wide, delicate, exuding a red juice when cut.

Gills: ascending, adnate, flesh colour with dark reddish edge (best seen

Spore-print: white.

Habitat: in grassy places, amongst bracken, often in vicinity of pines. Autumnal. Common.

The small red-brown cap, red-dotted gill edge and presence of a red juice in the stem are distinctive. *M. haematopus* also produces a red juice from the cut stem but produces clustered fruitbodies on wood; the caps are dull brown with a wine-red tint and hoary bloom, while the stems are pale above and dark red-brown below. *M. crocata* attached to twigs amongst beech leaves is a striking species which exudes a bright orange juice.

MYCENA GALOPUS
Milking Mycena

Cap: 1 cm diam., broadly bell-shaped, pale with brown centre and radiating striae.

Stem: up to 5 cm high, 2 mm wide, greyish below, almost white above, when broken exuding a white milk.

Gills: adnexed, white.

Spore-print: white.

Habitat: woodland, hedge-bottoms, heaths etc. Autumnal. Very common.

The pale cap with brown radiating striae and presence of a white milk in the stem are distinctive characters. A pure white variant var. *candida* is fairly common. *M. leucogala* is distinguished by its more conical very dark brown to almost black cap and stem. No other *Mycena* species possesses a white milk. It should be noted that when the stems of old fruitbodies of *M. galopus* are broken, it may be difficult to demonstrate the presence of milk. Ideally this test should be applied to young specimens.

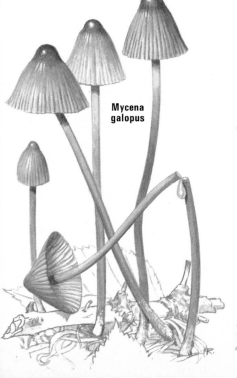

Mycena galopus

Mycena leucogala

51

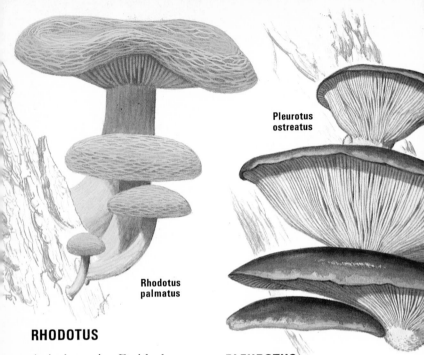

Pleurotus
ostreatus

Rhodotus
palmatus

RHODOTUS

A single species. Fruitbody
lignicolous, fleshy, pink to apricot
throughout. Cap with gelatinous
wrinkled surface. Stem central,
tough. Gills adnate or adnexed.
Spore-print salmon-pink.

RHODOTUS PALMATUS
Cap: 3·5-6 cm diam., convex, pink
at first then apricot; surface firm-
gelatinous, with raised wrinkled
meshwork.
Stem: 3-5 cm high, 5-10 mm wide,
central to slightly excentric, paler
than cap.
Gills: adnexed or adnate, deep,
distant, pinkish.
Flesh: same colour as cap.
Habitat: tufted on trunks or fallen
branches of elm. Autumnal but persisting into early winter. Occasional
but more common than formerly
owing to abundance of dead elms.

The lignicolous habit, strikingly
beautiful pink colour and wrinkled
gelatinous surface make for easy
identification.

PLEUROTUS

About 6 lignicolous species
forming solitary or clustered,
tough, fleshy, shell-shaped
bracket-like fruitbodies with
excentric to lateral stems, often
very reduced or rudimentary.
Gills decurrent. Spore-print
white to lilac.

PLEUROTUS DRYINUS
Syn: *P. corticatus*
Cap: up to 15 cm wide, solitary,
shell-shaped or bracket-like, whitish
covered with pale-greyish felt disrupting into small rather indistinct
scales especially toward the stem.
Stem: 3 cm long, 2-3 cm wide, short,
lateral to very excentric, often ascending, white with an indistinct ring or
ring-zone.
Gills: decurrent, distant, white.
Habitat: on live trunks of deciduous
trees. Autumnal. Occasional.

The large, shell-shaped fruitbody

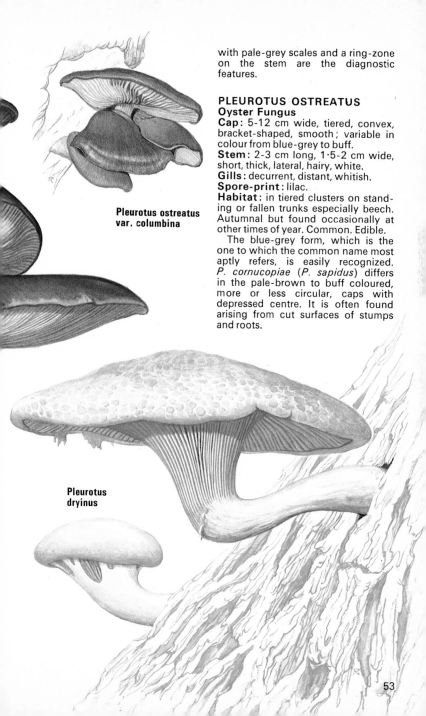

with pale-grey scales and a ring-zone on the stem are the diagnostic features.

PLEUROTUS OSTREATUS
Oyster Fungus
Cap: 5-12 cm wide, tiered, convex, bracket-shaped, smooth; variable in colour from blue-grey to buff.
Stem: 2-3 cm long, 1·5-2 cm wide, short, thick, lateral, hairy, white.
Gills: decurrent, distant, whitish.
Spore-print: lilac.
Habitat: in tiered clusters on standing or fallen trunks especially beech. Autumnal but found occasionally at other times of year. Common. Edible.

The blue-grey form, which is the one to which the common name most aptly refers, is easily recognized. *P. cornucopiae* (*P. sapidus*) differs in the pale-brown to buff coloured, more or less circular, caps with depressed centre. It is often found arising from cut surfaces of stumps and roots.

Pleurotus ostreatus var. columbina

Pleurotus dryinus

53

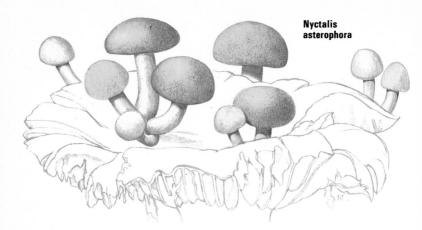

Nyctalis
asterophora

NYCTALIS
Pick-a-back Fungi

Two species growing parasitically
on certain Russulas. Fruitbodies
small and hemispherical with
greyish cap or with cap surface
brown and powdery.

NYCTALIS ASTEROPHORA
Syn: *Asterophora lycoperdoides*
Cap: 1·5-2 cm diam., hemispherical,
surface brown, powdery.
Stem: 1-1·5 cm high, 2 mm wide,
white.
Gills: lacking.
Habitat: clustered on rotting speci-
mens of *Russula nigricans.* Autumnal.
Rare.

Distinguished from *N. parasitica*
by the powdery cap and lack of gills.

NYCTALIS PARASITICA
Syn: *Asterophora parasitica*
Cap: 1-1·5 cm diam., bell-shaped,
flat or concave, lilac-grey.
Stem: 1·5-3 cm high, 1-2 mm wide,
white.
Gills: adnate with decurrent tooth,
thick, distant, often distorted, white
becoming brown due to production
of chlamydospores.
Habitat: clustered on various rotting
Russulas. Autumnal. Rare.

N. asterophora differs in having
a brown powdery surface to the cap
and complete lack of gills.

Nyctalis
parasitica

VOLVARIELLA

Ten species ranging in size from small to robust. Cap fleshy, either dry and silky fibrillose to smooth and sticky, usually white, sometimes grey or flushed yellowish. Stem lacking a ring but with a well formed volva.

VOLVARIELLA BOMBYCINA
Syn: *Volvaria bombycina*
Cap: 6-8 cm high, conical or bell-shaped, entirely and densely covered with silky, hair-like, often upturned fibrils, white to pale-yellow.
Stem: 7-10 cm high, 1 cm wide, white.
Volva: large, sac-like, sheathing, with conspicuous free limb, dark-brown and blotched on outer surface.
Gills: free, crowded, becoming pink.
Spore-print: pink.
Habitat: on wood, generally inside hollow elm trunks. Autumnal. Rare.

VOLVARIELLA SPECIOSA
Syn: *Volvaria speciosa; V. gliocephala*
Cap: 7-12 cm diam., at first conical, then convex with central boss, smooth, sticky, dirty-white with grey-brown centre.
Stem: 9-13 cm high, 1-1·5 cm wide, tall, narrowing above, fragile, whitish.
Volva: whitish, sac-like, but free limb relatively short and tending to collapse onto base of stem.
Gills: free, crowded, deep, white, then slowly pink with age.
Habitat: grassland, richly composted soil, near heaps of rotting grass or straw. Autumn. Occasional. Edible.

The large size, sticky pale cap with brown centre and presence of a volva are the diagnostic features.

There are several small white species of which *V. surrecta* is distinctive in growing parasitically on *Clitocybe nebularis.*

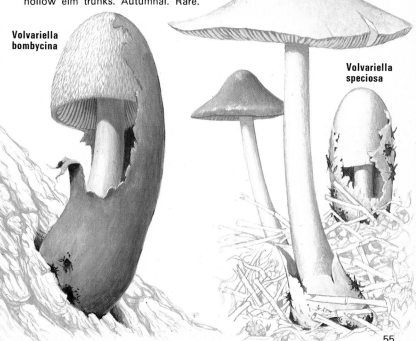

Volvariella bombycina

Volvariella speciosa

CLITOPILUS

Four species ranging from small and delicate with thin caps to fleshy, robust, fruitbodies with strongly decurrent gills. Spore-print pink.

CLITOPILUS PRUNULUS
The Miller
Cap: 3-6 cm diam., convex, then flat, top-shaped in section, smooth, dry, white to dirty-white sometimes flushed grey.
Stem: 3-5 cm high, 1-1·2 cm wide, short, thick, whitish.
Gills: decurrent, whitish at first, finally pinkish-flesh-colour.
Smell: strongly of meal, especially when bruised.
Spore-print: pink.
Habitat: deciduous woodland. Occasional. Edible.

PLUTEUS

A genus of over 30 species, both lignicolous and terrestrial. Caps are smooth or fibrillose, and may appear indistinctly scaly at centre. They may also be brightly coloured in shades of yellow, orange, or greenish but are more commonly dull, brownish or greyish. Spore-print pink.

PLUTEUS CERVINUS
Cap: 6-12 cm diam., strongly convex, umber to sooty-brown, often streaky.
Stem: 6-10 cm high, 9-12 mm wide, white, lower portion speckled with dark brown fibrils.
Gills: free, crowded, rather deep, white, eventually pinkish flesh colour.
Spore-print: pink.
Habitat: stumps and sawdust heaps, often several together. Autumnal but also sporadically throughout the year. Common. Edible.

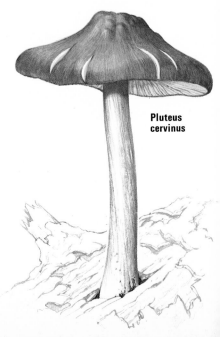

Pluteus
cervinus

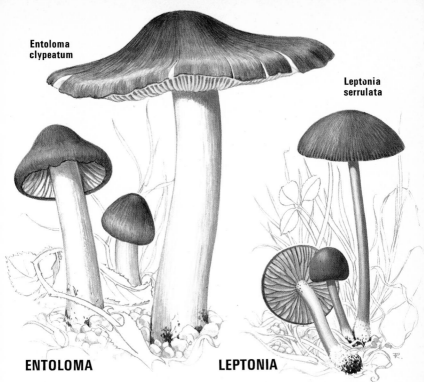

Entoloma clypeatum

Leptonia serrulata

ENTOLOMA

In the restricted sense a genus of about 25 species. Traditionally Entolomas are rather robust, fleshy, terrestrial fungi with pink sinuate gills and a pink spore-print.

ENTOLOMA CLYPEATUM
Syn : *Rhodophyllus clypeatus*
Cap: 3-6 cm diam., bell-shaped, then shallowly bell-shaped with central boss, grey-brown with darker radial streaks, drying paler.
Stem: 4-6 cm high, 6-15 mm wide, dirty-white to greyish with fibrillose surface.
Gills: sinuate, deep, distant, greyish becoming pink, edge irregularly wavy.
Flesh: greyish when water-soaked, white when dry, smelling of meal when crushed.
Spore-print: pink.
Habitat: with rosaceous shrubs and trees, hawthorn, fruit trees. Vernal.

LEPTONIA

About 50 species. The genus comprises small, mostly terrestrial, species which occur mainly in short turf. The caps, seldom more than 3 cm diam., are usually convex with a depressed umbilicate centre and many have a scurfy-fibrillose or felty surface; the stems are usually long and slender.

LEPTONIA SERRULATA
Cap: 2·5 cm diam., convex, often dimpled at centre, blue-black with fibrillose surface, becoming smoky-brown with age.
Stem: 4-5 cm high, 2 mm wide, tall, slender, steely blue usually with blue-black dots at apex.
Gills: at first blue-grey, becoming pinkish but with a black-dotted edge.
Habitat: short turf. Autumnal. Occasional.

57

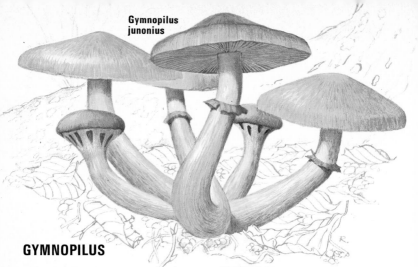

Gymnopilus junonius

GYMNOPILUS

Nine species with golden-yellow to rich tawny caps.

GYMNOPILUS PENETRANS
Syn : *Flammula penetrans*
Cap: 3-5 cm diam., broadly bell-shaped to flattened with a central boss, smooth, golden-tawny.
Stem: 4-5·5 cm high, 3-5 mm wide, striate, fibrous, yellow above, same colour as cap below with white base, ring lacking.
Gills: adnate to slightly decurrent, yellow with rusty spots, finally tawny.
Flesh: yellow in cap, rusty-brown in stem.
Spore-print: rusty-brown.
Habitat: on twigs in coniferous woodland. Autumnal. Very common.

GYMNOPILUS JUNONIUS
Syn : *Pholiota spectabilis*
Cap: 6-12 cm diam., convex, fleshy, bright tawny or golden yellow; surface fibrillose or disrupting into indistinct fibrillose scales.
Stem: 7-15 cm high, 1·2-3 cm wide, tapering at base, fibrillose, same colour as cap but paler, and with membranous ring near apex.
Ring: yellowish, soon collapsing back onto stem.
Gills: adnate, crowded, sometimes with decurrent tooth, yellow becoming rust-coloured.
Flesh: pale yellowish.
Spore-print: rusty-brown.
Habitat: forming dense tufts at the base of trunks (sometimes living) or on stumps of deciduous trees. Autumnal. Common.

Gymnopilus penetrans

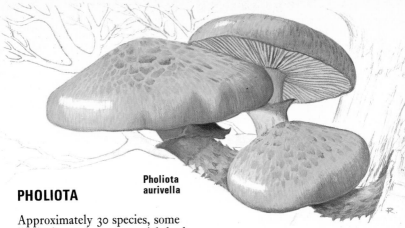

Pholiota aurivella

PHOLIOTA

Approximately 30 species, some lignicolous others terrestrial, both medium-sized and large, with smooth or scaly, dry or glutinous caps which are often bright-yellowish or tawny. Spore-print cigar-brown to rusty-brown.

PHOLIOTA AURIVELLA
Cap: 8-15 cm diam., shallowly convex, glutinous, deep yellow with rusty-brown centre, ornamented with conspicuous darker rusty-brown gelatinous scales.
Stem: 9-12 cm long, 1·5 cm wide, more or less horizontal, dry, fibrillose, yellowish becoming brownish below, with a quick fading fibrillose ring or ring-zone, below which it is often covered with pale hairy fibrils or small recurved fibrillose scales.
Gills: adnate or sinuate, deep, pale yellow then rusty-brown.
Flesh: pale yellow, brown in base of stem.
Spore-print: rusty brown.
Habitat: in small clusters high up on the trunks of deciduous trees, especially beech, sometimes on fallen trunks. Autumnal. Occasional.

PHOLIOTA SQUARROSA
The Shaggy Pholiota
Cap: 6-8 cm diam., convex, dry, pale ochre, entirely covered with prominent densely crowded up-turned bristly scales giving a coarse shaggy appearance.
Stem: 6-10 cm high, 1-1·5 cm wide, same colour as cap and covered with similar recurved scales below the fibrillose ring.
Gills: adnate with decurrent tooth, crowded, at first yellowish then pale rust-coloured.
Flesh: pale.
Spore-print: rusty-brown.
Habitat: parasitic on many species of deciduous trees, forming dense shaggy tufts' at the base of living trunks. Autumnal. Fairly common.

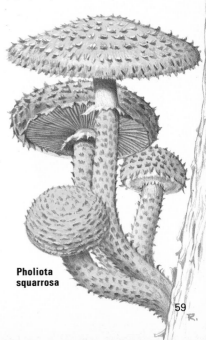

Pholiota squarrosa

59

GALERINA

About 25 species, both terrestrial and lignicolous. Caps seldom more than 3·5 cm diam., convex to broadly bell-shaped, brown and conspicuously striate.

GALERINA MUTABILIS
Syn: *Pholiota mutabilis Kuehneromyces mutabilis*
Cap: 2-3·5 cm diam., convex to broadly bell-shaped, watery date brown with striate margin when moist but drying out conspicuously from centre to tan colour. Fruitbodies when semi-dry are sharply two-coloured with tan centre and an abrupt, broad, watery-brown marginal zone.
Stem: 3-5·5 cm high, 3·5-5 mm wide, pale yellowish above becoming dark-brown and covered with paler recurved scales below the ring.
Gills: adnate with decurrent tooth, pale then cinnamon-brown.
Spore-print: cinnamon-brown.
Habitat: tufted on stumps of broadleaved trees. Autumnal. Fairly common. Edible.

AGROCYBE

Twelve medium to small terrestrial or lignicolous species with smooth, convex to flat, creamy-ochre, tan to dull brown caps. Stem fairly stout to slender, with or without a ring. Gills adnate to adnexed, clay brown. Spore-print dull brown.

AGROCYBE PRAECOX
Syn: *Pholiota praecox*
Cap: 3-5 cm diam., convex, smooth, cream-coloured with ochre-coloured flush at centre.
Stem: 5-7 cm high, 5-7 mm wide, tall, slender, white, with ring.
Ring: membranous, whitish.
Gills: adnate, clay-brown, crowded.
Flesh: white in cap, brown in stem, smelling of meal when crushed.
Spore-print: clay brown.
Habitat: in grassy places, roadside verges. Vernal. Common.

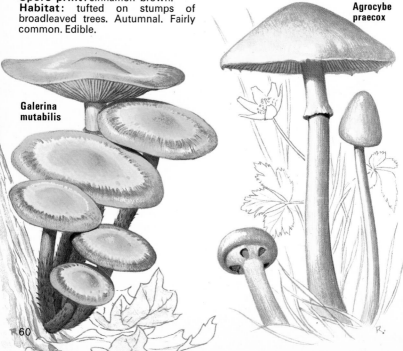

Agrocybe praecox

Galerina mutabilis

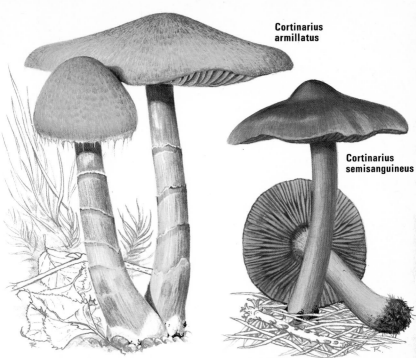

Cortinarius armillatus

Cortinarius semisanguineus

CORTINARIUS

Our largest single genus comprising about 300 species having in common a rusty spore-print and a cobwebby cortina (ring–zone) on the stem.

CORTINARIUS ARMILLATUS
Cap: 6-10 cm diam., shallowly convex, brick-colour to tan, smooth to indistinctly fibrillose.
Stem: 7-12 cm high, 7-12 mm wide, tall, fibrillose, with bulbous base, pale-brown with striking brick-red bands below.
Gills: adnate, cinnamon then rusty-brown.
Spore-print: rusty-brown.
Habitat: on heaths and commonland with birch trees. Autumnal. Occasional.

A distinctive species on account of its size and red banding at the base of the stem.

CORTINARIUS SEMISANGUINEUS
Cap: 3-6 cm diam., shallowly convex to flat, minutely fibrillose, brown or tawny-brown tinged olive.
Stem: 4-8 cm high, 3-6 mm wide, slender, yellowish-brown often with olive tints, ring-zone yellowish.
Gills: adnate, blood red.
Flesh: yellowish brown.
Spore-print: rusty-brown.
Habitat: gregarious in coniferous woodland. Autumnal. Occasional.

Cortinarius sanguineus, which occurs in similar habitats, differs in being entirely dark blood-red.

CORTINARIUS PSEUDOSALOR
Cap: 3-5 cm diam., broadly conical to campanulate with grooved wrinkled margin, glutinous, dull violet brown with olive tint.
Stem: 5-7 cm high, 1-1·5 cm wide, tapering below, dry and white above, but covered with violet gluten from just above the mid point to the base.

Gills: adnate, crowded, becoming rusty-brown but with pale violet edge.
Spore-print: rusty-brown.
Habitat: deciduous woodland, especially with beech. Autumnal. Very common.

Recognized by the broadly conical glutinous cap, stem covered below in violet gluten and the lilac edge to the gills.

TUBARIA

Five species with flattened cap and broadly adnate to subdecurrent brown gills, and growing either on twigs or on the ground.

TUBARIA FURFURACEA
Cap: 2·5-4 cm diam., convex then flat, cinnamon-brown when moist, drying out to pale biscuit colour from centre, but margin long remaining brown, striate and water-soaked, often with a broken ring of pale tufted scales near the edge.
Stem: 3·5-5 cm, same colour as cap but paler, sometimes with rather indistinct ring-zone.
Gills: very broadly adnate to sub-decurrent, cinnamon-brown.
Spore-print: yellowish-brown.
Habitat: on twigs in deciduous woodland, or in hedge-bottoms. Autumnal, but persisting into winter and occurring sporadically throughout the year. Very common.

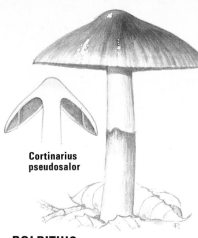

Cortinarius
pseudosalor

BOLBITIUS

A single species. Fruitbody tall, delicate, with thin flat slimy fluted cap, bright yellow at centre, brown and watery at margin. Gills free or adnexed.

Bolbitius
vitellinus

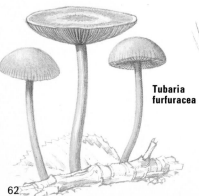

Tubaria
furfuracea

BOLBITIUS VITELLINUS

Cap: 2·5-3·5 cm diam., at first acorn-shaped, becoming flat with strongly grooved margin, surface smooth, sticky, chrome yellow at first, but this colour later restricted to centre with the margin brown and watery.

Stem: 6-10 cm high, 3-5 mm wide; pale yellow to whitish with hoary surface.

Gills: free to adnexed, cinnamon-brown.

Spore-print: rusty-brown.

Habitat: solitary or gregarious in richly manured grass and amongst wood chips. Autumnal. Occasional.

INOCYBE

A large genus of about 86 terrestrial, small to medium-sized species, most of which cannot be identified without a microscope.

INOCYBE ASTEROSPORA

Cap: 3-4·5 cm diam., shallowly convex with central boss, fawn-brown to chestnut at centre but toward margin surface disrupting into radial fibrils, same colour as cap, on a pale background.

Stem: 3·5-6 cm high, 5 mm wide, cylindrical with conspicuous basal bulb having a well-defined margin; surface striate, entirely pruinose, fawn-brown, paler at apex.

Gills: adnexed, becoming cinnamon to clay-brown.

Habitat: deciduous woodland. Autumnal. Occasional. POISONOUS.

Recognized by its relatively tall, brown stem with conspicuous white basal bulb and brown, radially fissured cap.

INOCYBE GEOPHYLLA

Cap: 1-2·5 cm diam., conical or bell-shaped, with white, silky fibrillose surface.

Stem: 2-3·5 cm high, 3-4 mm wide, white, pruinose at apex.

Gills: adnexed, white, slowly becoming clay-brown.

Spore-print: clay-brown.

Habitat: deciduous woodland. Autumnal. Common. POISONOUS.

INOCYBE GEOPHYLLA LILACINA

As above but entirely lilac except sometimes for a yellowish tint at centre, and for the gills ultimately becoming clay-brown.

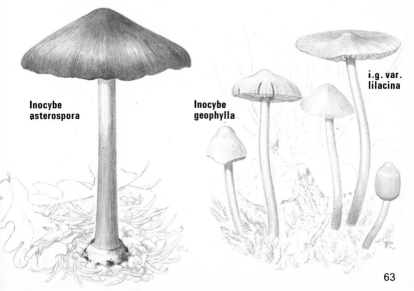

Inocybe asterospora

Inocybe geophylla

i.g. var. lilacina

HEBELOMA

About 20 terrestrial species, mostly fleshy, varying from small to large, usually with smooth, pale biscuit–coloured caps often with a flush of brown or tan at centre and with a sticky surface.

HEBELOMA CRUSTULINIFORME
Fairy Cake Fungus
Cap: 3-5·5 cm diam., convex, pale biscuit-coloured with flush of tan at centre, surface sticky.
Stem: 3·5-5 cm high, 5-8 mm wide, white, pruinose above.
Gills: sinuate becoming clay-brown, often with watery droplets along the edge in damp weather.
Flesh: white, smelling strongly of raddish.
Spore-print: clay-brown.
Habitat: deciduous woodland. Autumnal. Fairly common. POISONOUS.

PAXILLUS

Four species, some terrestrial others lignicolous.

PAXILLUS INVOLUTUS
The Roll Rim
Cap: 5-11 cm diam., convex with strongly enrolled margin, becoming shallowly depressed at centre, surface glutinous in wet weather, yellowish-brown, smooth except toward the edge which is often ribbed and downy.
Stem: 6-7 cm high, 1-1·2 cm wide, central, same colour as cap but paler, often streaky.
Gills: decurrent, yellowish-brown becoming dark red-brown when bruised.
Spore-print: ochre-brown.
Habitat: heathy places, associated with birch. Autumnal. Very common. POISONOUS.

Easily recognized by its association with birch, the brown glutinous cap with enrolled, woolly, ribbed margin.

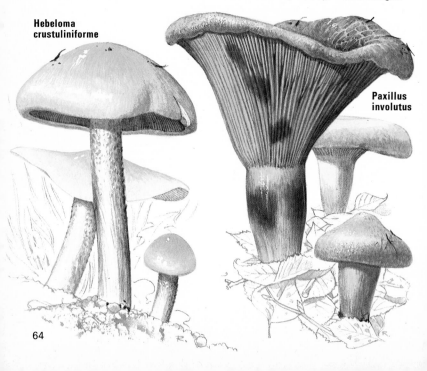

Hebeloma crustuliniforme

Paxillus involutus

CREPIDOTUS

A genus of 18 species, growing on wood, herbaceous plant material, rarely on soil. Cap shell-shaped ranging from about 1 cm diam. to 5 cm diam., mostly white or whitish, occasionally watery brown, sometimes with a separable gelatinous skin. Spore-print pinkish-brown to snuff-brown.

The small white species of *Pleurotus* or *Pleurotellus* are at once distinguished by having a white print.

CREPIDOTUS MOLLIS

Cap: 2-5 cm diam., horizontal, shell-shaped, growing in tiers; attached directly to wood, soft, flabby, pale watery yellowish-brown with striate margin, drying paler, surface gelatinous, separable as an elastic skin.

Spore-print: snuff-brown.

Habitat: tiered on stumps. Autumn. Occasional.

Distinguished by the shell-shaped tiered brackets with horizontal brown gills and a separable gelatinous cuticle. The latter is readily demonstrated if the cap is stretched laterally, when the transparent, gelatinous layer will be seen as an elastic film between the gills. *C. calolepis* has the cap ornamented with bright rusty-brown, fibrillose scales. The many white species can only be distinguished after microscopic examination.

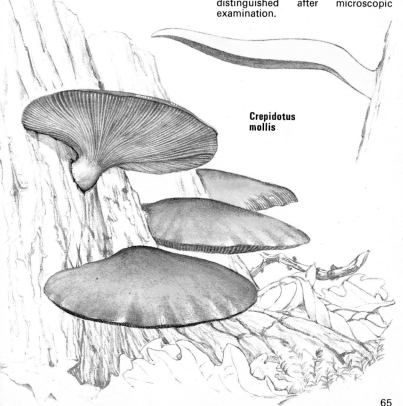

Crepidotus mollis

COPRINUS
Ink Caps

About 90 species, large or minute, mostly growing on dung or richly manured ground, sometimes at the base of stumps or from roots and then often tufted. Cap surface usually white or grey, more rarely brown or some other colour, either powdery like meal or with fibrillose scales. Stem with or without a ring. Gills black, usually liquefying after a few hours giving an ink-like fluid.

COPRINUS PLICATILIS
The Little Japanese Umbrella
Cap: 2-3 cm diam., acorn-shaped, yellowish-brown and closely striate at first, then flat, coarsely grooved or fluted, grey with small depressed tan-coloured disc, very thin, short-lived, almost translucent.
Stem: 6-8 cm high, 3 mm wide, whitish, very delicate.
Gills: free, attached to a collar around stem apex, scarcely liquefying.
Spore-print: black.
Habitat: damp grass, lawns, roadside verges. Autumnal. Common.

COPRINUS COMATUS
Shaggy Ink Cap; Lawyers Wig
Cap: 6-14 cm high, cylindrical, opening slightly at base, eventually bell-shaped, white, with buff-coloured central disc, surface broken up into shaggy scales, margin closely striate becoming greyish, finally black on maturity. Entire cap gradually dripping away from margin as an inky fluid.
Stem: up to 30 cm high, 1-1·5 cm wide, white with movable membranous ring toward base.
Gills: free, crowded, white, near stem apex, then pink and finally black at margin.
Spore-print: black.
Habitat: gregarious on rubbish tips, roadside verges, fields and gardens. Autumnal. Fairly common. Edible, provided the gills have not started to liquefy.

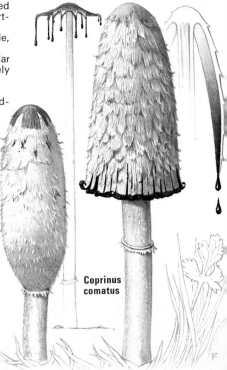

Coprinus
comatus

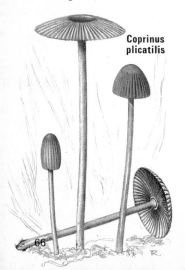

Coprinus
plicatilis

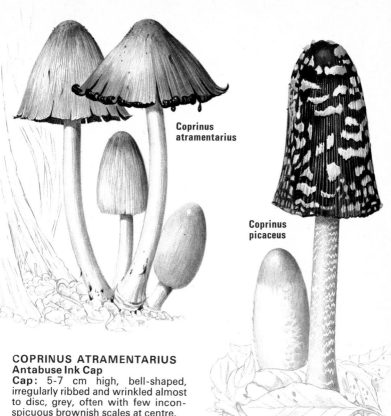

Coprinus atramentarius

Coprinus picaceus

COPRINUS ATRAMENTARIUS
Antabuse Ink Cap
Cap: 5-7 cm high, bell-shaped, irregularly ribbed and wrinkled almost to disc, grey, often with few inconspicuous brownish scales at centre.

Stem: 7-9 cm high, about 1 cm wide at base, where there is a conspicuous oblique ring-zone, white.

Gills: free, whitish, becoming grey, finally black, liquefying.

Spore-print: black.

Habitat: tufted in vicinity of stumps of deciduous trees or from roots, often in gardens, fields etc. Autumnal but sporadically throughout the year from early spring. Common. Edible, but causing sickness if eaten with alcohol due to a substance contained in the fungus identical with the drug Antabuse, which is used in treatment of chronic alcoholics.

This robust, tufted, greyish ink cap is readily identified. *C. acuminatus* also tufted or solitary, is a woodland fungus which is less robust and has a darker, more egg-shaped cap.

COPRINUS PICACEUS
The Magpie
Cap: 8 cm high, oval, then bell-shaped, cap surface disrupting into conspicuous white patches on a blackish background.

Stem: up to 25 cm high, 1 cm wide, hollow, fragile, white, lacking a ring.

Gills: free, crowded, white above, then pinkish, buff and finally black at margin, liquefying readily.

Spore-print: black.

Habitat: beech woods. Autumnal.

Easily identified by the robust stature and cap with contrasting white patches on a black background. This black and white colouring has led to the popular name 'The Magpie'.

Coprinus
disseminatus

AGARICUS

About 40 small, medium- or
large-sized, terrestrial species
with cap surface either white or
brown. Stem easily separable
from cap, and with membranous
ring. Ring simple or appearing
cog-wheel-like below. Gills free,
finally deep purple-brown. Flesh
unchanging, reddening or
becoming yellow when bruised.
Spore-print purple-brown or
chocolate-brown. Before eating,
it is essential to check that the
gills are purple-brown at
maturity, and that there is a ring
on the stem but no volva.

COPRINUS DISSEMINATUS
The Trooping Crumble Cap
Syn : *Psathyrella disseminata*
Cap: 5-10 mm high, acorn-shaped to
hemispherical, pale yellowish-clay
becoming greyish at the margin,
closely grooved almost to the centre,
minutely hairy under a very strong
lens.
Stem: 1-3·5 cm high, 1-1·5 mm
wide, white, very delicate and brittle.
Gills: adnate, dark grey to blackish,
scarcely liquefying.
Spore-print: black.
Habitat: densely gregarious, covering entire stumps in myriads of tiny,
brittle, bell-shaped, biscuit-coloured
fruitbodies. Autumnal, but also sporadically throughout the year. Common.

AGARICUS CAMPESTRIS
Field Mushroom
Syn : *Psalliota campestris*
Cap: 4·5-8 cm diam., convex then
flat, white, surface often disrupting
into indistinct fibrillose scales espec-
ially around centre.
Stem: 4-6 cm high, 1-1·5 cm wide,
short, squat, with pointed base and
ring.
Ring: poorly developed, simple,
often little more than a torn fringe.
Gills: free, at first pink then purplish-
brown.
Flesh: white, sometimes reddish in
stem when cut.
Spore-print: purplish-brown.
Habitat: grassy places, pastures,

Agaricus
campestris

often in fairy-rings. Autumnal. Occasional. Edible.

Distinctive characters are the squat growth form, pink colour of the young gills which finally become purple-brown and the poorly developed ring.

AGARICUS ARVENSIS
Horse Mushroom
Syn : *Psalliota arvensis*
Cap: 6-11 cm diam., hemispherical, becoming shallowly convex, white, often creamy with age, sometimes faintly yellow when handled, but never vividly so.
Stem: 8-12 cm high, 1·5-2 cm wide, tall, cylindrical, with bulbous base, white with membranous ring.
Ring: large, pendulous, high on stem, underside like radiating spokes of a wheel (cog-wheel-like).
Gills: white, then brownish, finally purplish-brown, never pink.
Flesh: white, unchanging.
Spore-print: purplish-brown.
Habitat: open grassland, hillsides, orchards etc, often in fairy-rings. Autumnal. Occasional. Edible.

AGARICUS BITORQUIS
Syn : *Psalliota bitorquis; Agaricus edulis*
Cap: 6-10 cm diam., shallowly convex with enrolled margin, whitish to pale buff.
Stem: 4-11 cm high, 2-3 cm wide, short, stout, sheathed below to an inferior ring.
Ring: double, the upper thicker and better developed, the lower narrower, thinner sometimes reminiscent of a volva, sometimes disrupting into bands of scales.
Gills: free; dirty-pink, then purple-brown.
Flesh: white, reddening slightly when broken.
Spore-print: purple-brown.
Habitat: often in soil around the base of trees in pavements, sometimes pushing up paving stones, and roadsides. Autumnal. Occasional. Edible.

The inferior double ring is diagnostic. Check that the spore-print is purple-brown. Specimens which stain bright yellow should be avoided (see *A. xanthodermus*).

Agaricus bitorquis

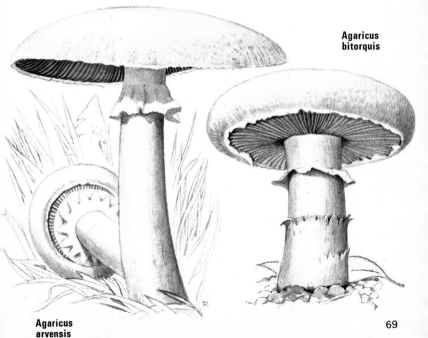

Agaricus arvensis

**Agaricus
xanthodermus**

**Agaricus
placomyces**

AGARICUS XANTHODERMUS
Yellow-Staining Mushroom
Syn : *Psalliota xanthoderma*
Cap: 5-8 cm diam., at first hemispherical with truncate top, later broadly bell-shaped to shallowly convex; white; often cracked or fissured toward margin; becoming vivid yellow when scratched.
Stem: 6-7 cm high, 1-1·5 cm wide; cylindrical with bulbous base; white, becoming yellow when scratched; ring present.
Ring: membranous.
Gills: whitish, then grey, finally purple-brown.
Flesh: white becoming vivid yellow in base of stem.
Spore-print: purple-brown.
Habitat: gregarious in pastures, woodland, gardens, shrubberies. Autumnal. Occasional. POISONOUS.

The white cap and yellow staining enable this species to be readily identified.

AGARICUS PLACOMYCES
Syn : *Psalliota placomyces; Agaricus meleagris*
Cap: 5-8 cm diam., shallowly convex with slightly depressed centre, surface disrupting except at disc into small, densely crowded, greyish-black scales on a whitish background; the disc is uniformly grey-black; on handling the cap bruises vivid yellow.
Stem: 7-9 cm high, 8-12 mm wide, rather tall and narrow, white, bruising yellow, with ring.
Gills: at first pink, then purplish-brown or blackish-brown.
Flesh: white, vivid, yellow in stem base.
Spore-print: purplish-brown.
Habitat: deciduous woodland. Autumnal. Occasional.

The blackish scaly cap and yellow staining which develops on bruising are distinctive characters. *Agaricus langei, A. haemorrhoidarius* and *A. silvaticus* are all species with brown scaly caps and reddening flesh.

STROPHARIA

About 15 terrestrial species of small to medium size, often with glutinous cap. Stem with membranous ring or prominent ring-zone. Gills sinuate or adnate. Spore-print black.

Species of *Agaricus* differ in having a dry cap, free gills and a purple-brown spore-print.

STROPHARIA AERUGINOSA
Verdigris Fungus
Cap: 3·5-5 cm diam., shallowly bell-shaped, glutinous, bright blue-green with tiny white fleck-like scales floating in the gluten toward the margin; with age the scales and gluten wash off leaving cap yellowish.
Stem: 4-6 cm high, 3-5 mm wide, with ring near apex, smooth and whitish above ring, pale blue-green and cottony-scaly below.

Ring: sometimes collapsing to blackish ring-zone indicated by trapped spores.
Gills: sinuate, smoky-brown with white edge.
Spore-print: black.
Habitat: deciduous woodland, gardens, often in nettle beds. Autumnal. Occasional.

Unmistakable due to the blue-green, glutinous cap.

STROPHARIA SEMIGLOBATA
Cap: 1-2·5 cm diam., hemispherical, pale yellowish to yellowish-tan, glutinous when wet, shiny when dry.
Stem: 4-8 cm high, 1-2 mm wide, tall, narrow, fragile, with black ring-zone near apex, dry above ring, glutinous below.
Gills: adnate, deep, dark-brown.
Habitat: especially on horse-droppings, richly manured places such as gardens and pastures. Autumnal. Locally common.

S. merdaria is similar but has a more brownish, flatter cap and an entirely dry stem.

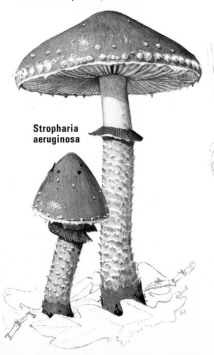

Stropharia aeruginosa

Stropharia semiglobata

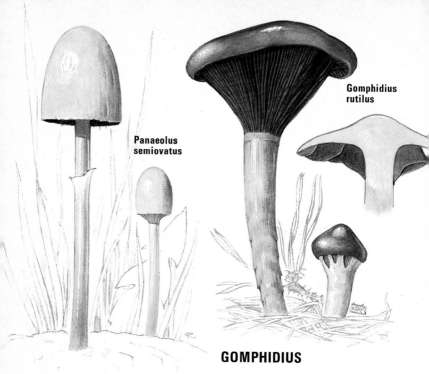

Panaeolus
semiovatus

Gomphidius
rutilus

PANAEOLUS

Twelve species, mostly small, with acorn-shaped caps and tall, delicate, narrow stems, growing on dung or in richly manured grassland.

PANAEOLUS SEMIOVATUS
Syn: *Anellaria semiovata; Panaeolus separatus; Anellaria separata*
Cap: 1·5-4 cm high, acorn- or egg-shaped, smooth, pale to greyish-buff, sticky when moist, strikingly shiny when dry.
Stem: 5-10 cm high, 3-5 mm wide, tall, fragile, with membranous ring near apex, whitish above, same colour as cap below ring.
Ring: membranous, soon collapsing.
Gills: adnate, black, mottled.
Spore-print: black.
Habitat: on horse droppings. Autumnal. Occasional.

GOMPHIDIUS

Five terrestrial medium-sized species restricted to coniferous woodland. Spore-print blackish. Cap top-shaped, sometimes with central boss or nipple; dry, sticky or glutinous.

GOMPHIDIUS RUTILUS
Syn: *Gomphidius viscidus*
Cap: 4-8 cm diam., convex often with central nipple, smooth, sticky, brown with purplish or wine-red tints, margin enrolled and yellowish or coppery.
Stem: 6-10 cm high, 1-1·5 cm wide, tapering below, fibrillose, purplish brown like cap, with broad cottony, buff-coloured ring-zone on upper portion.
Gills: decurrent, distant, yellowish-grey finally dingy purplish.
Flesh: entirely yellowish-tan.
Spore-print: blackish.
Habitat: pine woods. Autumnal. Occasional.

PSATHYRELLA

A genus of over 60 species, both terrestrial and lignicolous, mostly small to medium-sized. Spore-print dark purplish-brown to almost black. Often very fragile, with caps commonly or broadly bell-shaped, sometimes flattened, either naked or with whitish or blackish scales, surface usually drying out rapidly and becoming very pale.

PSATHYRELLA CANDOLLEANA
Syn : *Hypholoma candolleanum*
Cap: 2·5-5 cm diam., shallowly bell-shaped to flat, pale creamy-ochre to whitish especially when dry, with tiny tooth-like remnants of veil hanging from margin when young.
Stem: 4-6 cm high, 4-5 mm wide, white, hollow, very brittle.
Gills: adnate, for a long time whitish then lilac-grey and finally brownish-black.
Spore-print: almost black.
Habitat: tufted on wood, stumps, roots, fence-posts. Spring to Autumn. Common.

The pale, flattened, tufted, fragile fruitbodies with glistening surface to the cap and lilac-grey gills are easy to identify. *P. hydrophila* is another densely tufted species, formerly more common than at present, with watery date-brown, striate caps, becoming opaque and pale tan on drying.

PSATHYRELLA GRACILIS
Cap: 1·5-2·5 cm diam., bell-shaped, dark brown or reddish-brown with striate margin when moist, opaque and pale biscuit colour when dry.
Stem: 8-10 cm high, 2 mm wide, very tall, fragile, pure white with white hairy fibrils at base.
Gills: adnate, blackish, with pink edge.
Habitat: solitary or gregarious, amongst grass or leaves, in deciduous woodland, hedge-bottoms, roadside verges. Autumn. Very common.

The tall, fragile white stem, small bell-shaped cap, often seen in the pale condition, with glistening surface, together with the dark gills edged with pink are the salient features.

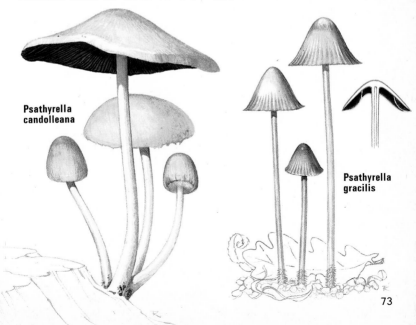

Psathyrella candolleana

Psathyrella gracilis

PSILOCYBE

About 20, mostly small terrestrial species of heaths and grassland, a few growing on twigs and sawdust. Gills adnexed or adnate, sometimes broadly so, reddish or purplish-brown.

PSILOCYBE SEMILANCEATA
Liberty Cap
Cap: 1-1·5 cm high, conical with sharp-pointed apex, sticky when moist, pale tan drying out to pale creamy-buff.
Stem: 3·5-5 cm high, 2 mm wide, paler than cap, sometimes blue at base.
Gills: adnate or adnexed; ascending; black with white edge.
Spore-print: purple-brown to almost black.
Habitat: often gregarious in grassland, lawns, fields, heaths. Autumnal. Fairly common. Hallucinogenic.

The characteristic shape of the tiny cap affords easy recognition.

LACRYMARIA

Two tufted species, apparently terrestrial, but growing from buried wood.

LACRYMARIA VELUTINA
Weeping Widow
Syn : *Hypholoma velutinum.*
Cap: 4-6 cm diam., bell-shaped or convex, yellowish-brown to clay-brown, densely radially fibrillose, with enrolled woolly fringed margin.
Stem: 6-7 cm high, 5-8 mm wide, same colour as, but paler than cap, fibrillose or scaly, with prominent ring-zone of whitish cottony fibrils often becoming black due to trapped spores.
Gills: adnexed or adnate, almost black, mottled, with white edge bearing watery droplets in damp weather.
Spore-print: almost black.
Habitat: tufted from roots or buried wood, in deciduous woodland. Spring to autumn. Common

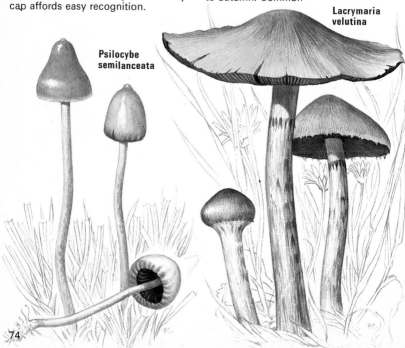

Lacrymaria velutina

Psilocybe semilanceata

HYPHOLOMA

Twelve species, both terrestrial and lignicolous, growing singly, gregariously or tufted.

HYPHOLOMA FASCICULARE
Sulphur Tuft
Syn: *Naematoloma fasciculare*
Cap: 2-4 cm diam., shallowly bell-shaped to shallowly convex, sulphur-yellow often with tawny flush at centre, margin with dark fibrillose remnants of veil.
Stem: 4-7 cm high, 4-8 mm wide, same colour as cap, sometimes flushed brown below, with poorly developed purplish-brown fibrillose ring-zone near apex.
Gills: sinuate, sulphur-yellow becoming olive.
Flesh: yellow with bitter taste.
Spore-print: purplish-brown.
Habitat: tufted at base of deciduous and coniferous tree-stumps. Autumnal, but sporadically throughout the year. Very common.

HYPHOLOMA SUBLATERITIUM
Syn: *Naematoloma sublateritium*
Cap: 4-7 cm diam., convex to flat, brick-red, paler and yellowish toward margin which often bears fibrillose remnants of veil.
Stem: 6-9 cm high, 8-10 mm wide, fibrillose, yellowish above reddish-brown below with fibrillose ring-zone near apex.
Gills: adnate or sinuate, yellowish becoming greyish-violet to purplish-brown.
Flesh: yellowish, but reddish-brown in base of stem, taste mild.
Spore-print: purplish-brown.
Habitat: tufted at base of deciduous tree-stumps. Autumnal. Occasional.

Recognized by brick-red cap and tufted habit at base of trunks or stumps. *H. fasciculare* differs in sulphur-yellow colour, olive gills and bitter taste. In addition the stem and flesh of *H. fasciculare* become bright orange with ammonia whereas in *H. sublateritium* the colour change is to bright yellow.

Hypholoma
fasciculare

Hypholoma
sublateritium

BOLETES

Fruitbodies are fleshy, with a cap and central stalk which may or may not bear a ring. The underside of the cap is sponge-like with tiny pores. These are the openings of densely crowded tubes lined with basidia which produce the spores.

Strobilomyces floccopus

SUILLUS

Fruitbodies with sticky cap, stem with or without a ring, sometimes with glandular dots at apex or throughout. Always with conifers.

SUILLUS GREVILLEI
Syn : *Boletus grevillei; Boletus elegans*
Cap: 5-9 cm diam., shallowly convex, often with slight central boss, then flat, glutinous, lemon-yellow to rusty-gold.
Stem: 6-8 cm high, 1-1·5 cm wide above, tapering to base, rather tall and narrow, brownish-yellow with whitish membranous ring near apex

Suillus grevillei

STROBILOMYCES

A single species with blackish cap covered in thick scales.

STROBILOMYCES FLOCCOPUS
Syn : *Strobilomyces strobilaceus*
Cap: 6-12 cm diam., convex, densely covered with thick cottony, pyramidal blackish scales with paler greyish in between.
Stem: 8-12 cm high, grey-black, tufted-scaly beneath ring-zone, pale and smooth at apex.
Pores: fairly large, greyish, bruising reddish.
Flesh: becoming reddish, finally black.
Habitat: deciduous woodland, especially beech. Autumnal. Rare.

which soon collapses leaving a broad ring-zone.
Pores: sulphur-yellow.
Flesh: pale yellow in cap, deeper in stem; sometimes with hint of lilac.
Habitat: always with larch. Autumnal. Occasional. Edible.

SUILLUS LUTEUS
Syn: *Boletus luteus*
Cap: 7-12 cm diam., bell-shaped, then flattened, often with slight central boss, glutinous when moist, dark chocolate- o purplish-brown, sometimes becoming rusty-tan with age.
Stem: 6-8 cm high, 1·7-2 cm wide, yellow with darker glandular dots above the well developed ring, whitish or pale brownish below.
Ring: spreading, membranous, white or greyish, often dark with age.
Pores: adnate to decurrent, dull yellow to deep ochre-yellow.
Flesh: whitish to pale lemon-yellow especially in stem.
Habitat: coniferous woodland. Autumnal. Fairly common. Edible.

SUILLUS BOVINUS
Jersey Cow Bolete
Syn: *Boletus bovinus*
Cap: 4-7 cm diam., convex at first then flattened, glutinous when moist, buff to pinkish-buff with pale margin.
Stem: 4·5-7 cm high, 6-9 mm wide, tapering downward, same colour as cap.
Ring: lacking.
Pores: decurrent, large, irregular, compound (each pore subdivided into smaller pores), dirty-yellow to rusty.
Flesh: yellowish or pinkish, reddish in stem.
Habitat: coniferous woodland. Autumnal. Common. Edible.

Recognized by its colour (resembling that of a Jersey cow) and large compound pores. *S. variegatus* has a dull yellow-brown cap flecked with small, dark-brown fibrillose scales, a rather stout, similarly-coloured stem lacking a ring and tiny dark cinnamon-brown pores.

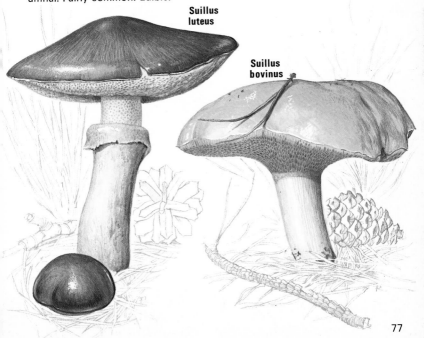

Suillus luteus

Suillus bovinus

Leccinum
scabrum

Leccinum
versipelle

LECCINUM

Fruitbody boletoid, with dry,
convex cap. Flesh often changing
colour when cut.

LECCINUM SCABRUM
Syn: *Boletus scaber*
Cap: 5-10 cm diam., convex to
shallowly convex, grey-brown with
an almost granular-mottled effect
under a lens.
Stem: 8-12 cm high, 2-3 cm wide,
tall, cylindrical, often enlarged below,
white but rough with conspicuous
black flocci.
Pores: almost free, minute, dingy
buff, bruising yellowish-brown.
Flesh: white, unchanging or faintly
pink.
Habitat: with birch, especially on
heaths. Autumnal. Very common.
Edible.

LECCINUM VERSIPELLE
Syn: *Boletus versipellis; B. testaceo-
scaber; L. testaceo-scabrum*
Cap: 6-13 cm diam., convex with
fringe-like margin which over-reaches
the pores, tawny-orange.
Stem: 9-14 cm high, 2·5-3 cm wide,
tall, cylindrical, slightly enlarged
below, white, rough-dotted with
black flocci.
Pores: almost free, minute, greyish.
Flesh: white, becoming greyish-
mauve, but blue-green at base of
stem, finally blackening.
Habitat: with birch, especially on
heaths. Autumnal. Common. Edible.

The bright tawny-orange cap and
black dotted stem make identification
easy. *L. aurantiacum* is similar but is
associated with poplar

BOLETUS

Fruitbody fleshy. Flesh often bruising indigo blue or blue-green when broken. Cap large to very large, convex, dry or humid but never conspicuously sticky, surface felty or downy, sometimes cracking. Stem thick, cylindrical or conspicuously bulbous, without a ring.

BOLETUS BADIUS
Syn: *Xerocomus badius*
Cap: 6-10 cm diam., shallowly convex, dark bay to chocolate-brown, slightly sticky in wet weather, shiny when dry but softly downy at margin.
Stem: 7-8 cm high, 1·5-2 cm wide, pale brown with darker streaks.
Pores: adnate or adnexed, small, cream to lemon-yellowish, turning blue-green when bruised.
Habitat: deciduous and coniferous woodland. Autumnal. Common. Edible.

Recognized by dark brown cap and pale creamy pores which rapidly become blue-green to touch.

BOLETUS CHRYSENTERON
Syn: *Xerocomus chrysenteron*
Cap: 5-7 cm diam., convex, madder-brown sometimes with olive tint, cracking in a chequered manner showing pale pinkish flesh in between; surface minutely felty.
Stem: 6-8 cm high, 1 cm wide, yellowish streaked red below.
Pores: adnate to adnexed, large angular, at first pale dull yellow then olive-yellow.
Flesh: pink under cuticle, yellowish elsewhere often reddish in stem, turning slightly blue when cut.
Habitat: deciduous woodland. Autumnal. Very common.

Recognized by madder-brown cap cracking to show pink flesh. *B. armeniacus* has a peach-coloured cap, while in *B. rubellus* it is blood-red, and in *B. pruinatus* dark purplish-bay to almost blackish with velvety look and with brighter yellow pores. In none of these species does the cap crack in such a conspicuously chequered manner.

B. subtomentosus, a common species of heathy areas, differs from *B. chrysenteron* in its velvety olive-tan coloured cap bruising dark brown on handling, the lack of pink colour under the cuticle which seldom cracks, a buff stem ornamented with coarse, branching, rusty ribs and large bright golden yellow pores, bluing slightly to touch.

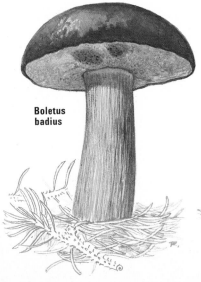

Boletus badius

Boletus chrysenteron

BOLETUS EDULIS
Penny Bun Bolete
Cap: 10-16 cm diam., strongly convex, usually chestnut brown.

Stem: 6-12 cm high, 2·5-4 cm wide, cylindrical but sometimes conspicuously swollen at base to 10 cm wide; pale with whitish network at least at apex.

Pores: adnexed, whitish becoming greenish-yellow.

Flesh: whitish unchanging, sometimes faintly pink in cap.

Habitat: deciduous and coniferous woodland. Late summer to autumn. Fairly common. Edible and excellent. (The 'Cep' or 'Steinpilz' of continental gourmets and a major constituent of many 'mushroom' soups).

The penny-bun coloured cap, whitish pores and pale stem with whitish net are distinctive. *B. aestivalis* (Syn : *B. reticulatus*) is separated by its pale straw to pale snuff-brown cap. *B. aereus* has blackish cap lacking red tints and *B. pinicola* red-brown cap.

BOLETUS ERYTHROPUS
Cap: 7-11 cm diam., convex, dry, minutely downy, dark bay-brown to red-brown.

Stem: 8-11 cm high, 1·5-2·5 cm wide, yellow densely covered with very minute red granular-dots.

Pores: free, minute, blood-red.

Flesh: yellow, instantly indigo-blue like rest of fruitbody when cut or bruised.

Habitat: deciduous woodland. Autumnal. Fairly common. Edible despite startling colour change of flesh.

Recognized by dark brown cap, blood-red pores, red-dotted stem and indigo blue colour when handled or in cut flesh. *B. luridus* is similar but has conspicuous red-network on stem. *B. queletii* is best distinguished by beetroot colour in the base of the stem.

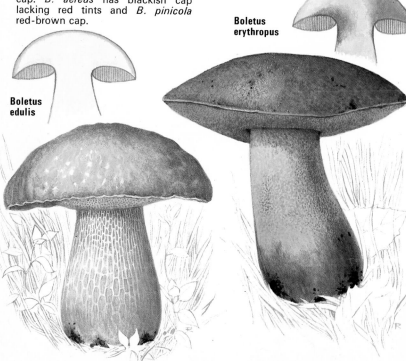

Boletus
erythropus

Boletus
edulis

BOLETUS PARASITICUS
Parasitic Bolete
Cap: 3-5 cm diam., convex then flat, minutely felty, tawny with olive tint.
Stem: 3-6 cm high, 6-8 mm wide, slender curved, yellowish streaked with rusty-red.
Pores: adnate, golden often with red blotches.
Flesh: yellow, reddish in stem, unchanging.
Habitat: one or more parasitic on the common Earth Ball (*Scleroderma citrinum*) in heathy or sandy localities. Autumnal. Rare.

Distinctive due to the unique habitat; the only parasitic bolete but in Japan *Xerocomus astereicola* grows on the earth star *Astraeus hygrometricus.*

Most boleti are liable to attack by the mould *Sepedonium chrysospermum,* which envelops them in a whitish mycelium and eventually produces a bright yellow powdery mass of spores, chiefly over the pores of the bolete.

BOLETUS SATANAS
The Devil's Boletus
Cap: 10-18 cm diam., convex with enrolled margin, pale-greyish.
Stem: 7-12 cm high, enormously swollen below up to 12 cm wide, yellow above, red below with conspicuous red network.
Pores: free, minute, blood-red.
Flesh: pale yellow, turns faintly blue in stem apex and over tubes.
Habitat: beechwoods on chalk. Autumnal. Rare. POISONOUS.

The large size, pale grey cap, enormously swollen stem with red net and the red pores are the salient features. In *B. satanoides* and *B. purpureus* the flesh becomes vivid blue when cut and the cap develops reddish or pinkish tones toward the margin. *B. calopus* differs in having yellow pores. Like *B. satanas* it has a red base to the stem but net is white or pinkish. It is the most common of the group of species discussed above.

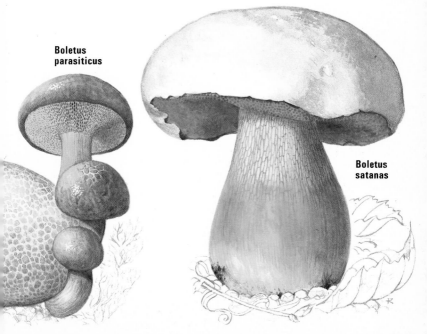

Boletus parasiticus

Boletus satanas

POLYPORES

These are the bracket fungi and are recognized by the poroid undersurface. The spores are produced by the basidia which line the tubes, the openings of which form the pores. Fruitbodies are usually lignicolous and either laterally stalked or in the form of rosette-like clusters, or solitary to densely-tiered, bracket-like fruitbodies commonly with woody or leathery texture.

COLTRICIA

Two species. Fruitbodies terrestrial, up to 10 cm diam.

COLTRICIA PERENNIS
Syn: *Polystictus perennis*
Cap: 2-10 cm diam., thin, leathery, funnel-shaped, adjacent fruitbodies often fused together, surface velvety, zoned in shades of tawny or rusty-brown but sometimes paling out to greyish-buff toward centre.
Stem: 3-5·5 cm high, 3-6 mm wide, velvety, rusty-brown.
Pores: small; lighter, brighter and more yellowish than stem, except when old, often with silky sheen.
Flesh: thin, fibrous, brown.
Spore-print: pale ochre.
Habitat: sandy heaths, sometimes on burnt ground. Autumnal but persisting for many months as discoloured blackened specimens.

GANODERMA

About 6 lignicolous species. Spore-print cocoa-brown.

GANODERMA APPLANATUM
Fruitbody: up to 30 cm diam. or more; flat, sessile, bracket-like, perennial, with thick, horny, dull crust, surface irregularly undulating, ornamented with conspicuous concentric grooves and varying in colour from buff to fawn or cocoa-brown; margin thin and rather acute.
Pores: very small, whitish, bruising brown.
Tubes: brown.
Flesh: fibrous, cinnamon-brown, thinner than tube layers.
Texture: very hard and woody.
Spore-print: cocoa-brown.
Habitat: parasitic on trunks and stumps, especially beech. Perennial. Very common.

Ganoderma applanatum showing the undersurface (right)

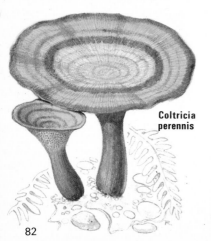

Coltricia
perennis

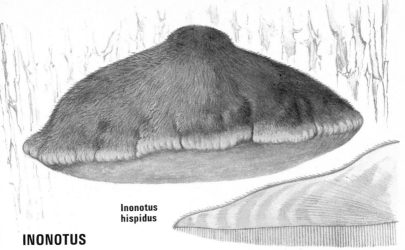

Inonotus
hispidus

INONOTUS

Six lignicolous species;
fruitbodies annual, solitary or in
tiers, with yellowish- to rusty-
brown fibrous flesh.

INONOTUS HISPIDUS
Syn : *Polyporus hispidus*
Fruitbody: 13-24 cm diam., bracket-
shaped, sessile, with bright yellow-
brown to rust-coloured velvety surface when young, but black and
bristly when old.
Pores: irregular, rusty-brown with
silvery sheen, glancing in the light,
often exuding watery droplets from
scattered pits in the surface.
Tubes: 3-4 cm deep.
Flesh: 5-6 cm thick, brown and
watery when actively growing, becoming dry in old specimens.
Spore-print: rusty.
Habitat: solitary on living deciduous
trees, especially ash, usually high on
the trunk, also on *Sophora japonica,*
apple, walnut, plane, elm. Spring and
summer. Common.

INONOTUS RADIATUS
Syn : *Polyporus radiatus*
Fruitbodies: 4-8 cm diam., sessile,
woody, tiered brackets, upper surface
minutely velvety becoming smooth,
tawny to rust-brown often with bright
yellowish margin, entirely black when
old.

Pores: minute, rusty-brown, glistening in the light.
Flesh: about 1 cm thick, woody,
fibrous, rusty-brown.
Spore-print: pale.
Habitat: tiered on stumps and
trunks of deciduous trees, especially
alder, occasionally on birch, but
rarely on other hosts. Summer to
autumn but long persistent in old
blackened condition. Fairly common.

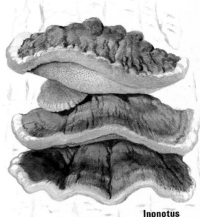

Inonotus
radiatus

83

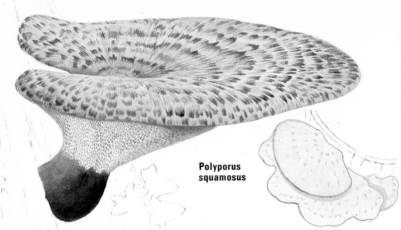

Polyporus
squamosus

POLYPORUS

About 8 lignicolous species; stem often black, at least at base. Caps varying from small to enormous (1-50 cm diam.) with smooth or scaly surface.

POLYPORUS SQUAMOSUS
Dryad's Saddle
Cap: 13-50 cm diam., fan-shaped, pale fawn with concentric rings of brown scales.
Stem: 3-8 cm high, 3-8 cm wide, relatively short, pale above with network due to rudimentary de-current pores, black and swollen below.
Pores: white to cream, very large, 1-3 mm diam.
Flesh: white, rubbery, up to 4 cm thick behind, smell mealy.
Spore-print: white.
Habitat: parasitic on trunks of deciduous trees fruiting on the living or dead host, particularly common on elm, beech and sycamore, often at considerable height above ground. Spring to summer. Frequent. Edible but worthless.

The enormous size of the brown-scaly cap, the black stem and large pores are diagnostic features.

Meripilus
giganteus

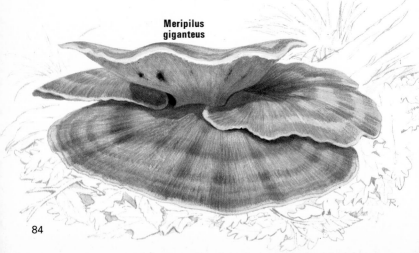

MERIPILUS

A single species forming large terrestrial rosette-like clusters at the base of trunks or stumps, or out in the open when arising from the roots. Flesh white. Spore-print white.

MERIPILUS GIGANTEUS
Syn: *Polyporus giganteus*
Fruitbodies: up to 60 cm diam., consisting of numerous clustered, fleshy, fan-shaped lobes, each with a yellowish, scurfy, granular-fibrillose surface ornamented with concentric brown zones; margin obtuse, fleshy.
Pores: late-forming; undersurface for a long time smooth, cream-coloured but bruising red-brown and finally black when handled.
Flesh: white, blackening when cut.
Habitat: forming large tufts at base of living or dead stumps or trunks, but sometimes growing from roots at some distance from trunks. Summer to autumn. Common.

GRIFOLA

Two species, both forming tufted clusters of caps or lobes at the base of trunks. Fruitbodies similar to those of *Meripilus*, but thinner and more numerous. In *G. umbellata* the fruitbody consists of a mass of branches each ending in a tiny umbrella-like cap about 1-4 cm across.

GRIFOLA FRONDOSA
Syn: *Polyporus frondosus;*
P. intybaceus
Fruitbody: up to 30 cm diam., consisting of a mass of small, thin, smoky-brown, often zoned, fan-shaped lobes, 3-7 cm across, each narrowed behind into a white stem-like portion attached to a common base.
Pores: decurrent, irregular, white.
Flesh: thin, white.
Spore-print: white.
Habitat: at base of living or dead stumps and trunks of deciduous trees. Summer to autumn. Occasional.

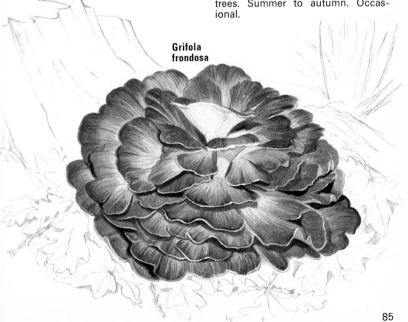

Grifola
frondosa

CORIOLUS

Two or three lignicolous species forming thin, annual, sessile, leathery, flexible, white-fleshed brackets with zoned, felty or hairy surface. The fruitbodies are often in conspicuous tiers. When pulled apart the broken surface appears fluffy.

CORIOLUS VERSICOLOR

Syn: *Polystictus versicolor; Trametes versicolor*
Fruitbodies: 3-8 cm diam., sometimes more when adjacent specimens merge together; thin, flexible sessile bracket-like, with felty surface ornamented with numerous colour zones varying from almost black or blue-black through smoky brown to tan or fawn especially at the margin. In addition to these colour zones there are often dark shining bands devoid of felt where the underlying surface shows through.
Pores: small, pale cream.
Flesh: white, about 2-3 mm thick.
Habitat: often conspicuously tiered, sometimes solitary, on stumps, trunks, fallen branches, bean poles or timber of both deciduous and coniferous trees throughout the year. Very common.

This is one of the commonest of the polypores and is readily identified by the felty surface with numerous colour zones.

Coriolus versicolor, one of the commonest of the polypores

CORIOLUS HIRSUTUS

Syn: *Polystictus hirsutus; Trametes hirsuta*
Fruitbodies: 7-11 cm diam., sessile, bracket-like with uniformly buff to fawn-coloured velvety-hirsute surface which is usually concentrically grooved.
Pores: small, whitish or cream with a distinct grey flush.
Flesh: white; up to 5 mm thick behind.
Habitat: usually on fallen beech trunks in exposed situations.

Coriolus hirsutus is thicker and firmer than C. versicolor

Bjerkandera adusta

HIRSCHIOPORUS

Two lignicolous species confined to coniferous wood. Fruitbodies thin, flexible, sessile brackets, the upper surface of which are greyish-buff and felty. The pore layer is bright violet and the pores either rounded or drawn out into tooth-like spines. Spore-print white.

HIRSCHIOPORUS ABIETINUS
Syn: *Polystictus abietinus; Trametes abietina*
Fruitbodies: 2-3·5 cm wide, thin, leathery, flexible brackets, often conspicuously tiered, upper surface concentrically grooved, felty, pale grey-buff.
Pores: small, irregular, vivid violet fading to flesh colour.
Flesh: thin, pale-brownish or purplish.
Habitat: fallen conifer trunks. All year round. Common.

BJERKANDERA

Two lignicolous species on deciduous trees. Fruitbodies thin, flexible, often tiered brackets with pale buff, minutely felty surface and small, grey to greyish pores.

BJERKANDERA ADUSTA
Syn: *Polyporus adustus*
Fruitbodies: 3-7 cm diam., thin, flexible, often tiered, brackets with upper surface minutely felty and uniformly buff to greyish-fawn.
Pores: small, ash-grey, darkening when dried.
Flesh: white.
Habitat: dead wood of deciduous trees, sometimes densely tiered and covering large areas. Throughout the year. Very common.

The small buff-coloured brackets with grey pores are distinctive. The latter feature separates it from *Coriolus versicolor* with which it often grows.

Hirschioporus abietinus

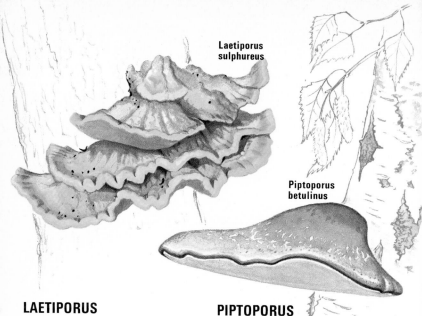

Laetiporus
sulphureus

Piptoporus
betulinus

LAETIPORUS

A single lignicolous species
producing clusters of thick,
fleshy, succulent orange-coloured
brackets with sulphur-yellow
margin. Flesh yellow at first but
soon whitish. Spore-print white.

LAETIPORUS SULPHUREUS
Syn : *Polyporus sulphureus*
Fruitbodies: up to 20 cm diam.,
consisting of thick, fleshy, tiered
brackets arising from a common base,
at first orange with obtuse sulphur-
yellow margin, fading to creamy-
white with age, surface smooth but
often irregularly undulating.
Pores: bright sulphur-yellow at
first, paling with age.
Flesh: yellowish and juicy at first,
soon whitish, developing a crumbly
texture.
Habitat: on trunks of living deciduous trees, especially oak, yew and
cherry but also on sweet chestnut
and willow. Spring to summer but
the old pale fruitbodies may persist
for several months. Occasional.

Instantly recognizable in the fresh
orange condition.

PIPTOPORUS

A single, large, lignicolous species
on birch forming smooth, hoof-
shaped, greyish or pale-brown
fruitbodies with a thin separable
skin. Flesh thick, white, rubbery.
Spore-print white.

PIPTOPORUS BETULINUS
Birch Bracket
Syn : *Polyporus betulinus*
Fruitbody: up to 20 cm diam., hoof-
shaped, occasionally with a basal
boss forming a short pseudo-stem,
surface covered by a greyish to pale-
brown or brown separable smooth
skin ; margin rounded.
Pores: very small, late-forming,
white.
Flesh: white, rubbery, up to 7 cm
thick behind.
Habitat: parasitic on birch, fruiting
on the living and dead trunks as
individual brackets although several
may be present at intervals up the
same tree. All the year round. Very
common.

The smooth, pale, hoof-shaped
brackets on birch are a familiar and
unmistakable sight.

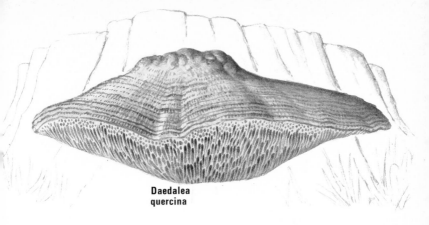

Daedalea
quercina

DAEDALEA

A single lignicolous species
virtually confined to oak.
Fruitbodies about 2 cm thick
forming tough, corky, flat
brackets.

DAEDALEA QUERCINA
The Maze Gill
Fruitbodies: up to 20 cm diam.,
woody to corky, bracket-like with
rough greyish-brown sometimes dull
ochre surface ornamented with concentric ridges and grooves, also with
radiating fibres and wrinkles.
Pores: pale, wood-coloured, radially
elongated and labyrinth-like, resembling a maze.
Flesh: pale wood-coloured, tough,
corky.
Habitat: singly or two or more
together, one immediately above the
other, on oak stumps. All the year
round. Common.
   The occurrence on oak and the
maze-like configuration of the pores
makes identification easy.

FISTULINA

A single lignicolous species
virtually confined to oak. Flesh
with appearance of raw beef-
steak and exuding a red juice.
Spore-print pale ochre.

FISTULINA HEPATICA
Beefsteak Fungus
Fruitbodies: up to 15 cm diam.,
solitary, fan-shaped with a narrow
point of attachment, sometimes with
a short stalk-like base, liver-coloured,
the surface roughened toward the
margin with minute warts which are
rudimentary tubes, otherwise smooth.
Tubes: individually separate.
Pores: yellowish-flesh-coloured.
Flesh: up to 5 cm thick behind with
colour, graining and texture of raw
beef-steak, also exuding a red juice;
strong acidic taste.
Habitat: on oak stumps or trunks,
usually near the ground. Autumnal.
Occasional. Edible but of poor
quality.

*The Beefsteak Fungus, Fistulina
hepatica*

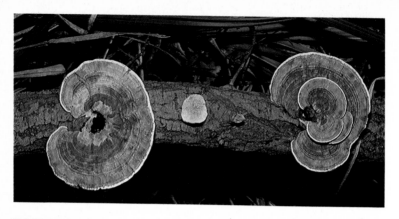

DAEDALEOPSIS

A single lignicolous species. Fruitbodies solitary or gregarious but not tiered. Surface zoned but smooth. Pores large, radially elongated, bruising red.

DAEDALEOPSIS CONFRAGOSA
Syn: *Trametes confragosa; Trametes rubescens*
Fruitbodies: up to 15 cm diam., flattened, shell-shaped, sometimes with a thickened basal hump at point of attachment, surface concentrically grooved and often irregularly radially wrinkled, zoned in lighter or darker shades of red-brown; margin acute, often white.

Daedaleopsis confragosa

Pores: radially elongated, slot-like, white to faintly grey, bruising red when rubbed if fruitbody is in active growth, and becoming lilac with a drop of ammonia; on old fruitbodies the pores become uniformly reddish-brown. Spore-print white.
Flesh: rubbery, zoned, whitish then reddish- or pale-brown.
Habitat: solitary or gregarious, several brackets often produced at intervals up a trunk or along branches of deciduous trees, especially willow. Throughout the year. Common.

The brackets usually occur on small trunks or branches and frequently appear to grasp them or encompass them.

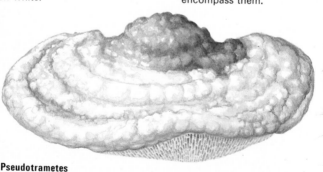

Pseudotrametes gibbosa

PSEUDOTRAMETES

A single lignicolous species forming thick pale brackets with concentrically grooved, minutely felty surface and an obtuse margin. Pores radially elongated and slot-like, not changing colour when bruised. Flesh white. Spore-print white.

PSEUDOTRAMETES GIBBOSA
Syn: *Trametes gibbosa*
Fruitbody: up to 20 cm diam., usually solitary, surface minutely felty, concentrically grooved, pale to greyish, but often greenish due to algal growth; margin rounded obtuse.
Pores: white, radially elongated, slot-like.
Flesh: white rubbery, up to 3 cm thick behind.
Habitat: stumps of deciduous trees, especially beech. Throughout the year. Very common.

The thick, pale, fruitbodies with unchanging slot-like pores are the salient features.

HETEROBASIDION

A single lignicolous species with hard woody perennial fruitbodies having a horny crust, indistinctly stratified tubes, cream-coloured flesh and a white spore-print.

HETEROBASIDION ANNOSUM
Syn: *Fomes annosus*
Fruitbodies: up to 20 cm diam., hard, woody, perennial, either entirely supine without any free-standing bracket, or with a bracket developed from the edge of the supine portion. Brackets often irregularly shaped with a very uneven knobbly surface ornamented with concentric grooves, reddish-brown with thin, white, acute margin.
Tubes: indistinctly stratified.
Pores: obvious to the eye, somewhat angular, cream.
Flesh: up to 1 cm thick, hard, cream or wood-coloured.
Habitat: essentially on roots of conifers but also on stumps. It is also found occasionally on roots of deciduous trees but in which case the brackets are poorly developed. *H. annosum* is an important parasite in conifer plantations, killing trees of various ages. All the year round. Common.

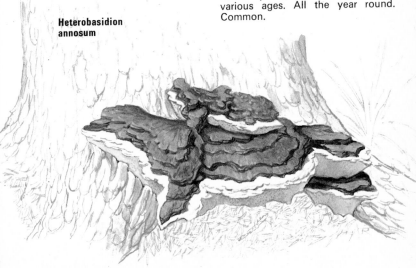

Heterobasidion annosum

HYDNOID FUNGI

Fruitbodies centrally or laterally stalked, rosette-like, bracket-shaped, forming coral-like clumps or supine, but with the lower fertile surface bearing crowded spines, hence the common name of Hedgehog fungi.

AURISCALPIUM

A single species growing from buried cones, with slender erect stalk bearing a tiny, kidney-shaped horizontal cap, the lower surface of which is densely spiny. Spore-print white.

AURISCALPIUM VULGARE
Ear Pick Fungus
Syn : *Hydnum vulgare*
Cap: 1-2 cm diam., horizontal, kidney-shaped, dark date-brown, velvety-hairy.
Stem: 3-5 cm high, 2 mm wide, velvety-hairy, same colour·as cap.
Spines: up to 3 mm long, brownish flesh colour, then blue-grey.
Habitat: on buried pine cones. Autumnal, rarely at other times of year. Rare.

HYDNUM

Two fleshy terrestrial species with smooth convex cap and short, stout central stem. Underside of cap bearing crowded white to pinkish spines. Flesh white. Spore-print white.

HYDNUM REPANDUM
Syn : *Dentinum repandum*
Cap: 5-10 cm diam., smooth, convex, fleshy, whitish to pinkish-buff.
Stem: 4-7 cm high, 1·5-2 cm wide, short, stout, whitish becoming yellowish at base and sometimes when bruised.
Spines: up to 6 mm long, whitish to salmon-pink, decurrent down top of stem.
Flesh: whitish, taste slightly bitter.
Habitat: deciduous woodland. Autumnal. Occasional. Edible and good.

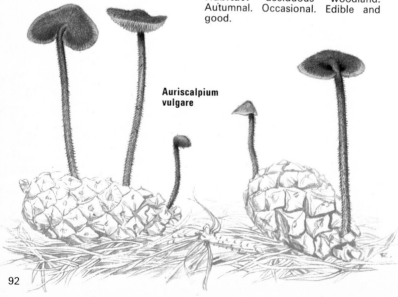

Auriscalpium
vulgare

Hydnum repandum

HERICIUM

Three lignicolous white species forming cushion-shaped fruit-bodies or much-branched coral-like masses bearing downward-pointing spines which may be short or up to 7 cm in length. Flesh white. Spore-print white.

HERICIUM ERINACEUM
Hedgehog Fungus
Syn: *Hydnum erinaceum*
Fruitbodies: 8-15 cm diam., comprising white cushion-shaped outgrowths bearing elongated downward-pointing spines up to 7 cm in length from the lower surface. Toward the margin of the upper surface there are often rudimentary spines. In section the flesh of the fruitbody shows a number of cavities.
Habitat: beech trunks. Autumnal. Rare.

The tuberculate fruitbody with very long spines is distinctive.

Hericium erinaceum, Hedgehog Fungus, growing on beech

93

CLAVARIOID FUNGI

The fruitbodies of the Fairy Clubs or Coral fungi, usually terrestrial or arising from herbaceous debris or twigs, vary from simple clubs to a densely-branched, coral-like structure. The fertile portion may be confined to the upper part of the club with the stem portion sterile, or the major part of the branched coral-like fruitbody may be covered in basidia which produce the spores.

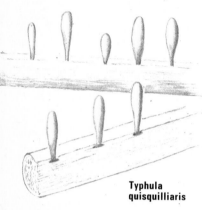

Typhula
quisquilliaris

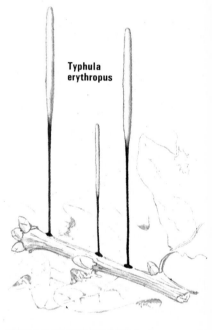

Typhula
erythropus

TYPHULA

About 8 or 9 species, some parasitic, on plant remains. Fruitbodies simple clubs varying from a few millimetres to several centimetres high, mostly white or pink, sometimes ochre-coloured, with delicate stalks.

TYPHULA QUISQUILLIARIS

Fruitbodies: up to 6 mm high, about 1 mm wide, small, club-shaped with no distinct stem, white, arising from a tiny brown, flattened, seed-like structure situated in the host tissue.
Habitat: dead bracken stems. Autumnal. Common.

This fungus is easily recognized since it forms a long row of tiny white, club-shaped fruitbodies along the length of the dead bracken stem. When this is split open the minute flattened brown seed-like structures can be seen just under the surface.

TYPHULA ERYTHROPUS

Fruitbodies: 2-3 cm high, about 1 mm wide, comprising a cylindrical white, club-shaped head occupying about a half to one-third of the total height.
Stem: dark reddish-brown to black, thread-like, arising from a small seed-like structure.
Habitat: deciduous woodland amongst litter, attached to leaves, stems, twigs. Autumn. Common.

The small white clubs and contrasting blackish thread-like stem are diagnostic.

CLAVARIADELPHUS

Four terrestrial woodland species. Fruitbodies simple clubs.

CLAVARIADELPHUS PISTILLARIS
Syn: *Clavaria pistillaris*
Fruitbodies: 10-30 cm high, 1-6 cm wide at apex, varying from cigar-like to club-shaped with distinct enlarged head, surface smooth but often longitudinally wrinkled below the swollen head; ochre-coloured becoming reddish or brownish especially toward the base.
Flesh: white, then purplish-brown where bruised.
Habitat: solitary or gregarious in beech woods on chalk. Autumnal. Rare. Edible.

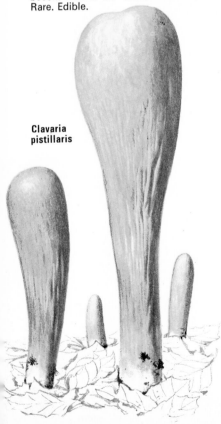

Clavaria
pistillaris

RAMARIA

Eight or nine species, with highly branched, coral-like, often brightly coloured and usually terrestrial fruitbodies. Spore-print pale yellow, ochre-coloured, cinnamon or rusty. Spores brown.

RAMARIA STRICTA
Syn: *Clavaria stricta*
Fruitbodies: up to 10 cm high, arising from a white mycelial felt, densely-branched, coral-like, fastigiate, at first pale ochre to pinkish-buff with the tips of the branchlets clear yellow, later brownish-ochre with similarly-coloured tips, all parts becoming darker or purplish when handled. Spores brown; minutely rough.
Habitat: on or near rotting stumps of deciduous trees. Autumnal. Rare.

The densely-branched, pale-ochre fruitbodies, which bruise purplish and occur on or in the vicinity of stumps, distinguish this fungus.

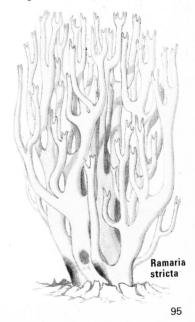

Ramaria
stricta

CLAVULINA

Four terrestrial woodland species with branched coral-like to simple fruitbodies ranging in colour from white or grey to lilac. Spore-print white.

CLAVULINA RUGOSA
Syn: *Clavaria rugosa*
Fruitbodies: up to 8 cm high, 5-10 mm wide, simple club-shaped but usually flattened and with longitudinal wrinkles, occasionally with one or more short antler-like side branches.
Habitat: deciduous woodland, often gregarious. Autumnal. Common.

CLAVULINA CINEREA
Syn: *Clavaria cinerea*
Fruitbodies: up to 9 cm high, irregularly-branched, coral-like and grey, with the ultimate branchlets rather blunt. The fruitbodies form dense clusters.
Habitat: deciduous woodland. Autumnal. Common.
　The dense, coral-like, grey clusters are characteristic.

Clavulina cristata

CLAVULINA CRISTATA
Syn: *Clavaria cristata*
Fruitbodies: 1·5-6 cm high, densely coral-like, white, with ultimate branches tending to be flattened and finely cristate.
Habitat: deciduous woodland. Autumnal. Common.
　The densely-tufted white, coral-like fruitbodies with cristate tips to the ultimate branches are easy to identify. However, specimens intermediate between this and *C. cinerea* may be difficult to place.

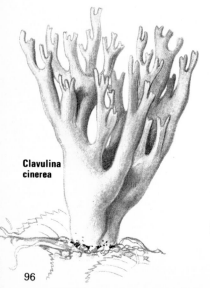

Clavulina cinerea

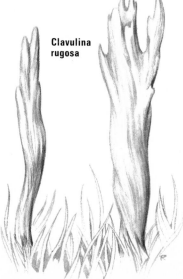

Clavulina rugosa

CLAVULINOPSIS

About 17 terrestrial species with simple or branched coral-like fruitbodies which are often yellow but also grey, brown or white in some species. Spore-print white to yellowish.

CLAVULINOPSIS CORNICULATA
Syn: *Clavaria corniculata*
Fruitbodies: up to 7 cm high, dichotomously branched, coral-like but branching rather lax with the ultimate tips incurved and crescentic, rather tough, varying from egg-yellow to ochre-yellow.
Flesh: same colour, with smell of meal when cut.
Habitat: often gregarious, in deciduous woodland or in open grassy situations. Autumnal. Occasional.

CLAVULINOPSIS FUSIFORMIS
Syn: *Clavaria fusiformis*
Fruitbodies: up to 10 cm high, comprising dense tufts of simple bright-yellow, somewhat flattened spindles with pointed tips, all arising from a common base.
Habitat: amongst grass on sandy heaths. Autumnal. Occasional.
The densely tufted habit is distinctive amongst all the other un-branched yellow species.

CLAVULINOPSIS HELVOLA
Syn: *Clavaria helvola; Clavaria inaqualis* of some authors
Fruitbodies: 2-4 cm high, 2-3 mm wide, simple club-shaped with obtuse apex, narrowed below into an indistinct sterile stem, bright orange-yellow.
Habitat: gregarious or in small groups amongst grass or moss on lawns, or in deciduous woodland. Autumnal. Common.
One of the commonest clavarioid fungi but easily confused with several other rarer species unless the spores are checked.

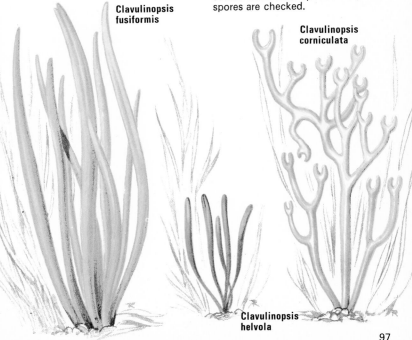

Clavulinopsis fusiformis

Clavulinopsis corniculata

Clavulinopsis helvola

CLAVARIA

About 15 simple, terrestrial species, varying in colour from white through grey to pink or purple, but pale citron-yellow in one instance. Spore-print white.

Clavaria argillacea

Clavaria vermicularis

CLAVARIA ARGILLACEA
Fruitbodies: 3-5 cm high, 3-6 mm wide, simple club-shaped, pale citron-yellow, with a distinct darker yellow stem.
Habitat: gregarious amongst short moss on heaths. Autumnal. Occasional.

CLAVARIA VERMICULARIS
Fruitbodies: up to 11 cm high, comprising densely-tufted, simple, hollow, very brittle, white, spindle-shaped fruitbodies each 4-5 mm wide, with a rather pointed apex.
Habitat: open grassy situations. Autumnal. Rare.

The densely-tufted, spindle-shaped, white fruitbodies are easily recognized although it may be difficult to distinguish certain collections from *C. fumosa* which has an identical growth form but in which the fruitbodies are typically smoky-grey.

Thelephora terrestris (left), clearly showing the small brown rosettes with soft, felty surface

Opposite page, Sparassis crispa. The cauliflower-like appearance makes it one of the easiest species to identify

STEREOID AND THELEPHOROID FUNGI

Fruitbodies either terrestrial or lignicolous. Very variable in shape, for example, bracket-like, coral-like with flattened branches, rosette-like, but with a smooth fertile surface devoid of gills, pores or spines, at most wrinkled.

SPARASSIS

Three species: *S. crispa* with its cauliflower-like appearance due to dense crowded curly lobes; *S. laminosa* with broader strap-like lobes and more open lax branching; and *S. simplex* which has single, erect, stalked palmate lobes arising from a fertile supine portion. Spore-print pale ochre.

SPARASSIS CRISPA
Cauliflower Fungus
Fruitbody: up to 30 cm diam., forming a pale-buff, cauliflower-like mass of innumerable tiny curled lobes, the whole mass arising from a short, white stem-like base.
Flesh: white with pleasant sweetish smell.
Habitat: at base of coniferous trunks. Autumnal. Fairly common. Edible.

THELEPHORA

About 8 or 9 terrestrial species. Fruitbodies erect, coral-like with flattened branches, rosette-like or supine, sometimes encrusting living plants. Spore-print dark brown.

THELEPHORA TERRESTRIS
Syn: *Phylacteria terrestris*
Fruitbodies: 2-6 cm diam., forming small irregular rosettes, either ascending or closely pressed to the ground, dark chocolate-brown, with soft, spongy, felt-like surface marked with radiating fibrils. Undersurface somewhat wrinkled and ornamented with minute warts, cocoa-brown.
Habitat: on the ground in coniferous woodland or on open sandy heaths. Autumnal. Common.
  The small brown rosettes with soft felty surface are characteristic.

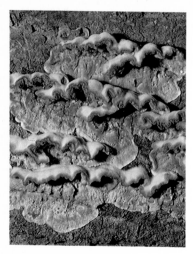

Chondrostereum purpureum

CHONDROSTEREUM

A single lignicolous species with thin, flexible, bracket-like fruitbodies in which the upper surface is pale and felt-like and the fertile surface smooth and purplish.

CHONDROSTEREUM PURPUREUM
Silver Leaf Fungus
Syn: *Stereum purpureum*
Fruitbodies: 2-4 cm diam., thin, flexible leathery brackets, often arising from a supine area, sometimes forming entirely supine patches. Upper surface felty with one or more concentric grooves, pale greyish-buff, often with dark line at or just in from the wavy margin. Undersurface bright lilac to purplish when in active growth, fading to brownish with age.
Habitat: saprophytic on a wide range of deciduous trees (stumps, branches etc) or parasitic especially on rosaceous trees and shrubs. When parasitic it sometimes causes silvering of the foliage as in Silver Leaf Disease of plum. All the year round. Common.

STEREUM

Seven lignicolous species with bracket-like fruitbodies, often arising from supine areas, sometimes entirely supine. Upper surface felty, lower surface smooth, sometimes exuding red juice when cut. Spore-print white.

STEREUM HIRSUTUM
Fruitbodies: 3-6 diam., thin, leathery, bracket-like with undulating margin, often tiered and frequently arising from supine patches. Upper surface coarsely felty-hairy, somewhat zoned, yellowish buff-brown to buff. Undersurface bright yellow when in active growth, becoming brownish to grey-brown with age.
Habitat: saprophytic on stumps, fallen trunks or branches of deciduous trees, especially beech. All the year round. Very common.
   Recognized by the coarsely hairy yellowish-brown brackets with bright yellow underside which often cover a considerable area. *S. subtomentosum* differs in the lower surface being paler and bruising yellow where chewed.

Stereum hirsutum

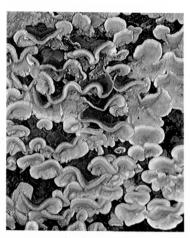

Stereum
gausapatum

STEREUM GAUSAPATUM
Fruitbodies: 2-3 cm diam., brackets similar to those of *S. hirsutum* but smaller and seldom produced in such abundance. Surface felty, zoned in shades of rusty-brown, grey-brown or fawn with wavy whitish (or pallid) margin. Underside dingy-brown, reddening when cut or scratched.
Habitat: particularly on oak, fallen branches, stumps. All the year round. Common.

Recognized by the well-formed brackets on oak which bleed when scratched. *S. sanguinolentum* is a similar bleeding species found only on conifer stumps and fallen branches. Here the brackets are rather thin, narrow and shelf-like with a rather silky-shining brown surface and pale greyish- to dirty-buff underside.

STEREUM RUGOSUM
Fruitbodies: forming extensive creamy supine patches on undersides of branches, only occasionally developing a very narrow, rigid, shelf-like portion, with dark-brown surface which may be slightly felty or naked. When scratched the cream-coloured surface reddens, and if broken the fruitbody is seen to have a stratified appearance under a lens.
Habitat: on small trunks, undersides of dead fallen or still attached branches of deciduous trees, especially hazel. All the year round. Common.

The creamy patches which redden when scratched are distinctive but if the fruitbody is dry when collected it may be necessary to moisten it before the red colour will develop.

Stereum
rugosum

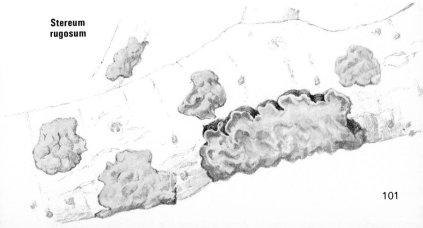

HYMENOCHAETE

About 6 lignicolous species, with bracket-like or supine fruitbodies which are some shade of brown. Flesh brown. Spore-print white.

HYMENOCHAETE RUBIGINOSA
Fruitbodies: 4-7 cm diam., densely-tiered, bracket-like, concentrically ridged and grooved, dark-brown to almost black and smooth, but when young with a minute rusty, velvety bloom. Lower surface chocolate-brown appearing as if waxed, with the extreme margin rust-brown or sometimes creamy-yellow. Flesh brown.
Habitat: old oak stumps. All the year round. Common.

The densely crowded dark brown to blackish brackets on oak stumps are easily identified, especially as despite being thin they are rigid and brittle. *H. tabacina* is a rare species, with bracket-like fruitbodies which are produced on fallen branches.

Dry Rot Fungus, Serpula lacrymans, growing on the wall of a derelict house

SERPULA

Two lignicolous species producing thick, soft, spongy brackets or supine patches. Both have the fertile surface ornamented with rusty-brown, grey-brown or olive-tinted 'pores'. Spore-print rust-brown.

SERPULA LACRYMANS
Dry Rot Fungus
Syn: *Gyrophana lacrymans; Merulius lacrymans*
Fruitbodies: often in the form of extensive supine patches up to 20 cm diam., in vicinity of sheets of felty or cottony fungal growth which are whitish or greyish with lemon or pinkish patches. When fruiting on a vertical surface soft spongy greyish brackets are formed, up to 10 cm diam., sometimes with lilac tints. Fertile surface ornamented with large, angular shallow pores, at first yellowish then bright rusty-brown.
Spore-print: rust-brown.
Habitat: always inside buildings, fruiting on damp woodwork, especially in cellars, but sometimes bursting out from plastered walls. All the year round. Common.

Hymenochaete rubiginosa

CONIOPHORA

Five or six species with lignicolous, supine, brown fruitbodies having a smooth to slightly knobbly surface. Spore-print olive-brown or brown.

Coniophora puteana, Wet Rot Fungus

CONIOPHORA PUTEANA
Wet Rot Fungus
Syn: *C. cerebella; C. laxa*
Fruitbodies: forming extensive supine patches with broad, white to cream-coloured radiating margin contrasting with the fertile areas which are often somewhat warted or knobby and at first creamy-ochre, then brown, frequently with olive tints.
Spore-print: olive-brown.
Habitat: spreading widely over dead stumps, trunks or timber. Throughout the year. Very common.

The supine fruitbodies with broad, white-fringed margin and contrasting olive-brown fertile areas are easily recognizable. Unlike *Serpula lacrymans* which once established can infect dry timber, *Coniophora puteana* needs constant dampness for spread to occur.

PHLEBIA

About 10 species. Originally restricted to supine species having the fertile surface ornamented with crowded, radiating wrinkles or short ridges. Spore-print white.

PHLEBIA MERISMOIDES
Syn: *P. radiata*
Fruitbodies: forming supine patches, 1·5-5 cm diam., which are strongly radiately wrinkled and purplish-flesh-coloured with a brilliant orange fringed margin, sometimes also with a lumpy orange centre. In old fruitbodies the orange colour may disappear.
Flesh: somewhat gelatinous.
Spore-print: white.
Habitat: on fallen trunks, especially beech. All the year round. Very common.

Phlebia merismoides

103

JELLY FUNGI

Fruitbodies mostly lignicolous, very variable in shape, but nearly always conspicuously gelatinous. When dry often very inconspicuous and horny or forming a varnish-like patch but swelling to previous gelatinous state when moistened.

Auricularia mesenterica

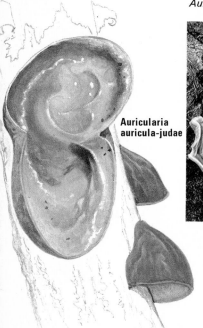

Auricularia
auricula-judae

AURICULARIA

Two lignicolous species with gelatinous, velvety-hairy fruitbodies which are either helmet-shaped or bracket-like. Spore-print white.

AURICULARIA AURICULA-JUDAE
Jew's Ear
Syn: *Hirneola auricular-judae; Auricularia auricula*
Fruitbodies: 3-6 cm diam., helmet-shaped, date-brown, firm-gelatinous, velvety. Undersurface with folds and ridges resembling the inside of an ear, pale purplish-brown.
Habitat: on branches of deciduous trees especially elder. All the year round. Very common.
Distinctive because of its shape, colour and gelatinous texture. A white form also occurs and should be reported if found.

AURICULARIA MESENTERICA
Fruitbodies: 4-7 cm diam., comprising densely-tiered, thick gelatinous brackets, with adjacent brackets becoming merged, surface velvety-hairy, zoned in shades of greyish-fawn and brown. Under surface reddish-purple with a whitish bloom, but when dry slate-grey to blackish, ornamented with radial folds especially near margin.
Habitat: covering large areas of old stumps with densely crowded brackets, especially elm. All through the year. Fairly common.
The gelatinous, velvety, zoned brackets are unmistakeable.

CALOCERA

Five lignicolous species forming branched or simple tough, gelatinous, club-shaped fruitbodies. Spore-print white to yellowish.

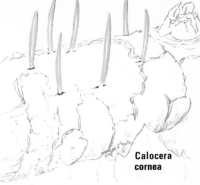

Calocera
cornea

CALOCERA CORNEA
Fruitbodies: 1-1·5 cm high, tough-gelatinous, forming simple, gregarious, pale yellowish, pointed 'clubs'.
Habitat: on fallen trunks and branches of deciduous trees, especially beech. Throughout the year. Common.

The simple, gregarious, pointed yellowish clubs on wood of deciduous trees are distinctive. *C. furcata* is very similar but occurs on conifer branches. *C. glossoides* has a longitudinally wrinkled head and distinct stalk. It occurs on fallen oak trunks.

The beautiful yellow tufts of Calocera viscosa

CALOCERA VISCOSA
Fruitbodies: 4-8 cm high, branched, club-shaped, bright egg-yellow, but of tough gelatinous texture; branches united below into a white rooting portion.
Habitat: on rotting conifer stumps. Autumnal. Common.

The bright yellow, branched, club-shaped fruitbodies on conifer stumps are distinctive. They are distinguished from the true Clavarias by their tough gelatinous texture, gliding easily between the fingers without breaking.

TREMELLA

About 7 or 8 species either growing on wood as conspicuous pendant fruitbodies formed of complicated brain-like or foliose masses, as wrinkled pustules or parasitically within the fruitbodies of other fungi (Polypores. *Aleurodiscus*, *Dacrymyces*). Spore-print white.

TREMELLA MESENTERICA
Fruitbodies: 1·5-5 cm diam., comprising pendant yellow gelatinous brain-like masses.
Habitat: on dead fallen or still attached branches of deciduous trees and shrubs. Throughout the year. Common.

TREMELLA FOLIACEA
Fruitbodies: up to 7 cm diam., comprising pendant, gelatinous, cinnamon- to red-brown, densely lobed, foliose masses.
Habitat: on dead attached or fallen branches of deciduous trees. Throughout the year. Occasional.

Tremella foliacea

Tremella mesenterica

Dacrymyces stillatus

EXIDIA

About 6 lignicolous species with gelatinous disc-shaped fruitbodies which are either distinctly stalked and pendulous or attached by a central point. Spore-print white.

EXIDIA GLANDULOSA
Witches' Butter
Fruitbodies: 1·5-4 cm diam., originating as pendulous, top-shaped gelatinous bodies, with flattened blackish discs ornamented with minute scattered papillae best seen under a lens. Sterile surface dark-brown to blackish, rough to touch.
Habitat: on fallen or still attached branches of deciduous trees. All the year round. Common.

DACRYMYCES

Ten lignicolous species with small, gelatinous, pustular or cup-shaped, yellowish or orange fruitbodies, seldom more than a few millimetres across. Spore-print white or yellowish.

DACRYMYCES STILLATUS
Fruitbodies: 2-6 mm diam., pustular, yellowish or orange, with smooth or wrinkled surface.
Habitat: on cut surfaces of stumps, fallen branches of both conifers and deciduous trees, also very common on damp timber (fence posts). All the year round. Common.

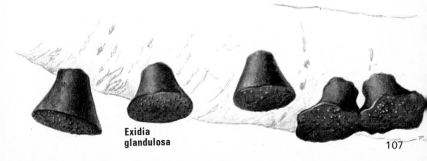

Exidia glandulosa

107

GASTEROMYCETES

These fungi include Puff Balls, Earth Stars, Stinkhorns and Bird's Nest fungi. They belong to a group known as the Phalloids, which in the young immature stage consist of an egg-like structure with a thick gelatinous wall, through which the receptacle emerges. This expands rapidly, reaching full size within an hour or so. The receptacle may consist of a spongy stalk with a fertile bell-shaped head or a tip bearing the brown glutinous spore mass. Alternatively the receptacle may have a floral or starfish like form or it may resemble a mesh-work hollow sphere. The fruitbodies are often pink or reddish and usually have a strong putrid smell, both features which serve to attract insects to disperse the glutinous olive-coloured spore mass.

PHALLUS

Two terrestrial species, but attached to buried wood or plant remains by whitish underground chord-like strands. The fruitbodies, remarkably phallic in appearance, consist of a stalk with a volva at its base and a pendulous, bell-shaped cap at its apex. The fragile stalk is hollow and has a sponge-like texture. The cap surface is covered by the glutinous, olive spore mass. Spore dispersal is by insects which are attracted by the strong putrid smell.

PHALLUS IMPUDICUS
Stinkhorn
Fruitbodies: 10-14 cm high, exceptionally up to 30 cm, comprising a fragile, white, spongy, hollow stalk with sac-like gelatinous remains of the egg forming a volva at its base, and supporting a pendulous bell-shaped cap at its apex.
Cap: white with reticulate surface which is only visible after removal of the glutinous, olive-coloured spore mass by insects.
Smell: extremely putrid, of rotting meat, so much so that the fungus is often smelled before it is seen.
Habitat: solitary or gregarious in deciduous woodland or gardens, attached by means of white strands to roots or buried wood. Summer to autumn. Very common. Edible in the egg stage.
The phallic appearance precludes all risk of confusion.

Stinkhorn, Phallus impudicus (left). The cap, which is white with a reticulate surface, only becomes visible after removal of the glutinous, olive-coloured spore mass by insects. The white, spongy stalk is hollow

MUTINUS

A single species consisting of a spongy, orange stem sheathed at its base by a white cylindrical gelatinous volva and having a fertile tip bearing the glutinous olive spore mass. Smell strong putrid. Spore dispersal is by insects.

MUTINUS CANINUS
Dog Phallus
Fruitbody: 7-9 cm high, consisting of a spongy, hollow, orange stem with a fertile tip covered by the glutinous, olive spore mass, and sheathed below in a white cylindrical gelatinous volval sac.
Smell: putrid.
Habitat: amongst fallen leaves in deciduous woodland. Autumnal. Occasional.
The shape and colour are distinctive.

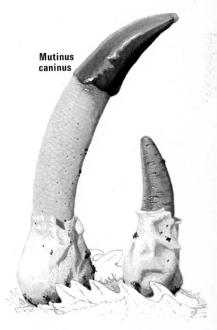

Mutinus
caninus

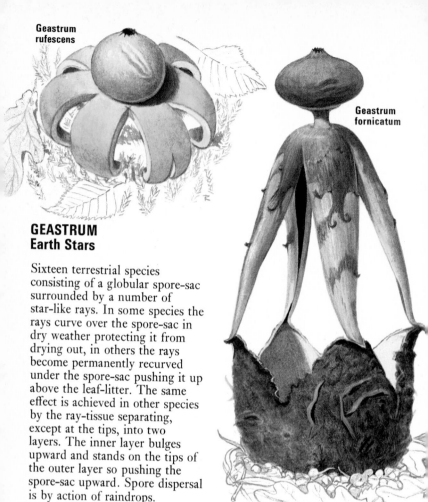

Geastrum rufescens

Geastrum fornicatum

GEASTRUM
Earth Stars

Sixteen terrestrial species consisting of a globular spore-sac surrounded by a number of star-like rays. In some species the rays curve over the spore-sac in dry weather protecting it from drying out, in others the rays become permanently recurved under the spore-sac pushing it up above the leaf-litter. The same effect is achieved in other species by the ray-tissue separating, except at the tips, into two layers. The inner layer bulges upward and stands on the tips of the outer layer so pushing the spore-sac upward. Spore dispersal is by action of raindrops.

GEASTRUM RUFESCENS
Syn: *G. fimbriatum*
Fruitbodies: 2·5-5 cm diam., when expanded.
Spore-sac: 1·5-2 cm diam., sessile, globular, nestling in a hollow at the centre of rays, smooth, pale-brown with an apical mouth consisting of a pore with fringed edge.
Rays: 6-8, flat or curved under spore sac, fleshy, yellowish to a very pale brown.
Habitat: in deciduous woodland. Autumnal. Occasional.

GEASTRUM FORNICATUM
Fruitbodies: 4-8 cm diam., 7-10 cm high when mature.
Spore-sac: 1-1·5 cm diam., smooth, globe-shaped, brownish, with fringed apical pore and a shortly stalked base.
Rays: 4 in number, the inner layer of the rays separating from the outer and curling so that they eventually stand on the tips of their counterpart, carrying the spore-sac clear of surrounding leaf-litter.
Habitat: deciduous woods. Autumnal. Rare.

A water droplet lands on Geastrum triplex, inducing it to expel a spore cloud

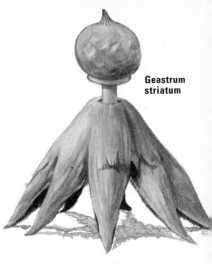

Geastrum striatum

GEASTRUM TRIPLEX

Fruitbodies: 6-10 cm diam., when expanded.

Spore-sac: up to 3 cm diam., smooth, sessile, globular, pale-brown, seated in a shallow cup formed from the base of the fleshy layer of the rays which breaks at that point when they curl under the fruitbody.

Rays: 4-8, thick, fleshy, yellowish to pale brown.

Habitat: deciduous woodland. Autumnal. Occasional.

Perhaps the commonest Earth Star, recognized by its large size, the shallow cup-like structure at the base of the spore-sac, and the fringed mouth surrounded by a halo.

GEASTRUM STRIATUM

Syn: *G. bryantii*

Fruitbodies: 3-6 cm diam., when expanded.

Spore-sac: about 1 cm diam., shortly stalked, flattened or sub-globose and flattened below, with a distinct rim or collar at base; bursts open through well-defined pore at top of a fluted cone, colour becoming almost blackish at maturity.

Rays: up to 8, becoming curled under fruitbody.

Habitat: deciduous woodland. Autumnal. Occasional.

Recognized by small size, dark flattened spore-sac with conical fluted mouth and collar at base.

Geastrum triplex

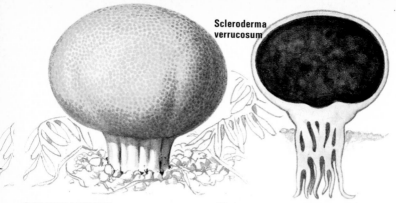

Scleroderma
verrucosum

SCLERODERMA

Six terrestrial species. Fruitbodies subglobose to somewhat
flattened, or pear-shaped with a
stalk-like base of varying
prominence, and a firm leathery
to thick rind.

SCLERODERMA VERRUCOSUM
Fruitbodies: 4-6 cm diam., up to
7 or 8 cm high, spherical or pear-
shaped, prolonged below into a
distinct stout stalk with ribbed
surface, and terminating in the soil as
a conspicuous mass of cottony
threads; yellowish-brown closely
covered over upper portion with
small brown scales. In section
this species is seen to have a thin
flexible tough rind and brownish-
black spore mass.
Habitat: on sandy heaths. Autumnal.
Fairly common.

SCLERODERMA CITRINUM
Earth Ball
Syn: *S. aurantium; S. vulgare*
Fruitbodies: 5-10 cm diam., hemispherical, often slightly flattened
above, with a basal chord-like
attachment or sometimes with a mass
of cottony threads, yellowish or
ochre-coloured with a conspicuously
rough, coarsely-scaly surface. When
cut through the fruitbody is seen to
have a thick whitish rind which often
flushes pink, surrounding a firm,
purplish-black spore mass which at
maturity becomes powdery.
Habitat: sandy heaths or woodland.
Summer to autumn. Very common.

Scleroderma
citrinum

The hemispherical shape, lack of a stalk, yellowish colour, scaly surface, thick rind and purple-black spore mass make this an easy fungus to identify. It is sometimes parasitized by *Boletus parasiticus* (see p. 81).

LANGERMANNIA

A single terrestrial species with spherical fruitbody of very large size lacking a well-defined sterile base and loosely attached to the soil by a tiny chord. The surface of the fruitbody flakes away to expose the powdery brown spores held in a cotton wool-like mass of threads.

LANGERMANNIA GIGANTEA
Syn: *C. caelata*

Fruitbodies: up to 30 cm diam., occasionally even larger, resembling a smooth white ball with kid-like texture to the surface. The interior is initially white and fleshy becoming yellowish, but as the spores mature the colour changes to olive-brown and the texture becomes cottony. Attachment to the soil is by means of a tiny chord, such that mature fruitbodies often become free and get blown by wind scattering spores in the process.

Habitat: on the ground, in fields, gardens and woodland. Autumnal. Occasional. Edible if eaten while the flesh is still white.

The large size and lack of sterile base are the distinctive characters.

Langermannia
gigantea

113

CALVATIA

Two terrestrial species with pear-shaped or pestle-shaped fruitbodies having a well-defined sterile base. The upper portion of the fruitbody flakes away to expose the powdery spores held in a mass of cottony threads.

CALVATIA UTRIFORMIS
Syn: *C. caelata*
Fruitbodies: up to 15 cm high, and of similar width, with more or less flattened top, narrowing slightly to a broad base. The surface is white and cracks into a regularly chequered pattern. At maturity the superficial whitish patches disappear exposing the underlying brown tissue. The interior of the fertile portion is at first white and fleshy but ultimately comes to contain the powdery olive-brown spores held in a mass of cottony threads. The sterile base, occupying about one-third of the volume of the fruitbody, is separated from the fertile area by a distinct membrane and has a sponge-like structure.
Habitat: open grassy places, heaths, sand dunes. Autumnal.

The shape of the fruitbody with chequered surface is characteristic.

CALVATIA EXCIPULIFORMIS
Syn: *Lycoperdon excipuliforme; Calvatia saccata*
Fruitbodies: up to 12 cm high, pestle-shaped with fertile head and well-developed sterile stalk. The surface is pale greyish-buff, densely covered over the upper portion with fine, scurfy whitish hair-like spines, interspersed with minute granular warts. In section the fertile head is seen to contain the powdery olive-brown spores held in a mass of cottony threads, while the sterile base has a whitish sponge-like appearance.
Habitat: pastures, heaths or deciduous woodland. Autumnal. Occasional.

This fungus is recognized by its large pestle-shaped fruitbodies with fine hair-like spines and minute granular warts, the presence of a distinct sterile stalk and by the flaking away of the top of the head.

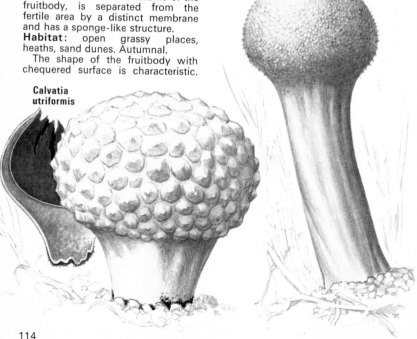

Calvatia
excipuliformis

Calvatia
utriformis

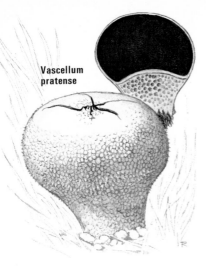

Vascellum pratense

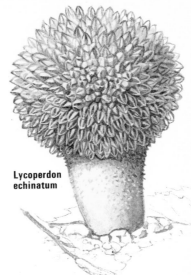

Lycoperdon echinatum

VASCELLUM

A single species with smallish, pear-shaped, terrestrial fruitbody differentiated into a fertile head and sterile stalk which are separated internally by a distinct diaphragm.

VASCELLUM PRATENSE
Syn: *Lycoperdon pratense; L. hiemale; L. depressum*
Fruitbodies: 2-4 cm diam., pear-shaped with fertile head and sterile stalk, white to cream, ornamented with scurfy white granules and small spines which may be united at the tips. At maturity the spines and granules disappear leaving a more or less smooth, pale brown shiny surface. In this stage a section through the fruitbody shows the fertile head with its dark olive-brown powdery spores held in a cottony mass of threads, separated by a distinct diaphragm from the sterile stalk with its sponge-like texture.
Habitat: in open situations amongst short turf and lawns, often forming fairy-rings. Summer to autumn. Very common.

LYCOPERDON

A genus of 7 or 8 terrestrial species with pear-shaped fruitbodies variously ornamented with granules, warts or spines. Internally the fruitbody is separated into a fertile head containing the powdery brown spores held in a mass of cottony threads and a sterile stalk with sponge-like structure.

LYCOPERDON ECHINATUM
Fruitbodies: 4-8 cm high, 4-6 cm diam., pear-shaped, densely covered with very conspicuous, long, curved, crowded spines approaching 5 mm in length, which are deciduous and leave a reticulate pattern on the surface of old specimens.
Stem: well-defined, covered with shorter spines and with a sponge-like interior.
Spore-mass: chocolate brown.
Habitat: deciduous woodland. Autumnal. Occasional.
   The long, curved, brown spines are distinctive.

115

Lycoperdon perlatum

LYCOPERDON PERLATUM
Syn : *L. gemmatum*
Fruitbodies: up to 8 cm high, and 5 cm diam., club-shaped with a distinct head and long cylindrical stem, white to pale-brown, densely covered above with prominent, but deciduous, crowded, white pyramidal warts each surrounded by a ring of minute granules. In old specimens from which the warts have disappeared the surface has a reticulate pattern formed by the minute rings of granules.
Stalk: with internal sponge-like structure.
Spore-mass: olive-brown.
Habitat: solitary, gregarious or even in small tufts in woodland. Autumnal. Very common.

The tallish, club-shaped fruitbodies with deciduous white pyramidal warts, which readily fall off during handling leaving a characteristic reticulate pattern on the surface, are easily distinguished.

LYCOPERDON PYRIFORME
Fruitbodies: up to 6 cm high, 3 cm wide, club-shaped or pear-shaped, often narrowly so, pale greyish or pale brownish, densely covered by fine scurfy granules.
Stalk: arising from conspicuous, white, chord-like strands; internal structure sponge-like but the cells rather small.
Spore-mass: greenish-yellow then olive-brown.
Habitat: gregarious, often in very large numbers on and around stumps of deciduous trees. Autumnal, but old weathered specimens can be found throughout the year. Very common.

Lycoperdon perlatum

BOVISTA

Lycoperdon pyriforme growing on an old stump

Six terrestrial species. Fruitbody globular without a stalk, entirely filled at maturity with the powdery spores held in a mass of cottony threads.

Bovista
nigrescens

BOVISTA NIGRESCENS

Fruitbodies: 3-6 cm diam., globular, at first white, but outer surface flaking away completely at maturity to expose the somewhat shiny blackish tissue beneath.

Stalk: lacking; fruitbodies with very tenuous attachment to the soil and usually becoming free. Old papery specimens persist for many months and may be blown for considerable distances scattering spores in the process.

Spore-mass: purplish-black.

Habitat: open grassland, dunes. Autumnal, but old weathered specimens may be found at any time.

The large black papery fruitbodies with broad mouth often marked by a recurved rim are easily recognized. *B. plumbea* is much smaller (1-2 cm diam.) and has a lead-grey colour after the outer white layer has flaked away.

117

BIRDS' NEST FUNGI

This group of Gasteromycetes is characterized by having fruitbodies resembling small cups or funnels which may or may not be fluted, and which usually have a felty or shaggy-hairy outer surface. The spores are produced in egg-like packages in the base of the fruitbodies which then resemble miniature birds' nests. Spore dispersal is by the splash of raindrops falling into the cups.

CRUCIBULUM

A single lignicolous species with small, barrel-shaped or deeply cup-shaped fruitbody containing numerous eggs, each attached to the interior by a fine thread.

CRUCIBULUM LAEVE
Syn: *C. vulgare*
Fruitbodies: 5-10 mm high, 5 mm wide, short, barrel-shaped or deeply cup-shaped, at first closed by thin membrane which soon ruptures. At maturity the cup has a bright cinnamon or yellowish felty surface and a smooth pale interior containing numerous pale-brown, disc-like eggs 1·5-2 mm diam., each attached by a thin thread to the wall.
Habitat: woody debris, twigs. Autumnal. Fairly common but often overlooked.

The small, parallel-sided, yellowish cup-like fruitbodies containing numerous eggs are easily recognizable.

Crucibulum laeve

Cyathus olla

CYATHUS

Three species either terrestrial or growing on dung or woody debris. Fruitbodies funnel-shaped, at first closed by a membrane which soon ruptures to reveal the eggs which are attached to the wall by a thin thread.

CYATHUS OLLA
Fruitbodies: 9-15 mm high, 7-13 mm wide, funnel-shaped with broad flaring mouth and narrow base, outer surface felty varying from pale greyish-buff to yellowish-grey-brown or brown; inner surface smooth, shiny, ranging from pale to dark-grey.
Eggs: about 2·5 mm, disc-shaped, either dark-grey or blackish.
Habitat: gregarious, often in large numbers on bare soil in woods and gardens, sometimes in flower pots. Spring to autumn. Occasional.

Recognized by the small flared funnel-shaped fruitbodies with felty surface and smooth grey interior, and the presence of relatively few but large dark-greyish eggs.

CYATHUS STRIATUS
Fruitbodies: 7-20 mm high, 7-14 mm wide, tall funnel-shaped, with shaggy-hairy, red-brown surface and shiny, fluted, greyish interior.
Eggs: about 1 mm diam., silvery-grey.
Habitat: gregarious, often in large numbers on woody debris. Autumnal. Occasional.

The tall, funnel-shaped, fluted fruitbodies with shaggy red-brown surface are readily distinguished. *C. stercoreus* is similar. This resembles *C. striatus* in having a shaggy-hirsute fruitbody, but the creamy-buff hairs are softer and more tangled and the interior of the funnel is non-fluted. In this particular collection the fruitbodies are small, ranging from 5-7 mm high and 3-4 mm wide. The identity of the fruitbodies, which commonly occur on dung, may be confirmed under the microscope by the presence of large spores.

Cyathus
striatus

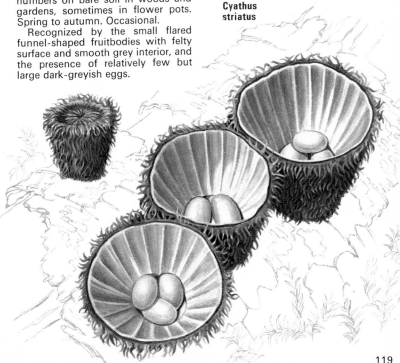

Glossary

adnate: (of gills) broadly attached to stem by their entire width (see page 13).
adnexed: (of gills) narrowly attached to stem by less than their entire width (see page 13).
adpressed: closely pressed to surface.

agarics: gill-bearing fleshy fungi, mushrooms.
apical: at the tip.
basidium (pl. basidia): spore-bearing organ of those fungi covered in this book.
campanulate: broadly bell-shaped.
cartilaginous: tough, firm, not easily snapped.
concolorous: of the same colour.
cortina: a zone of cobweb-like threads near the top of the stem in some agarics.
crispate: (of cap margin) wavy.
cristate: finely divided, fringed, crested.
deciduous: (of ring on stem) soon falling away.
decurrent: (of gills) running down the stem (see page 13).

dehiscence: (of fruitbodies) opening at maturity by a pore or splitting open to facilitate liberation of spores.
dentate: (of cap margin) fringed with tiny tooth-like fragments.
dichotomous: branching equally into two like a tuning fork.
fasciculate: in tufts or bundles.
fastigate: having the growth form of a Lombardy Poplar, i.e. with many densely-crowded, erect branches.
fibrillose: formed of threads or fibrils.
flocci: tiny, woolly scales.
glabrous: naked, smooth.
hispid: covered in stiff bristles or hairs.
hydrated: water-soaked.
membranous: thin, skin-like.
mycelium: sterile, felt-like or cobwebby vegetative stage of a fungus.
Papilla: small, nipple-like outgrowth.
pruinose: having a bloom like that of a ripe plum; as if covered with a fine layer of frost.
pustule: a tiny cushion-shaped outgrowth, hence pustular.
pyriform: pear-shaped, with a head which gradually narrows below into a stalk-like base.
rugulose: uneven, with fine, rather indistinct wrinkles.
saccate: sac-like or cup-shaped.
sessile: without a stalk.
sinuate: (of gills) notched just before joining stem (see page 13).
squamulose: bearing tiny scales.
stellate: star-shaped.
striate: marked with fine lines; (of cap margin) indicating that it is radially lined often due to the gills showing through.
supine: (of fruitbodies) prostrate, without a cap, forming paint-like patches on trunks or branches.
tomentose: having a felt-like texture.
tuberculate: bearing small warts.
veil: a sheath-like covering especially appertaining to the agaric fruitbody.
velar: derived from a veil.
viscid: sticky.
volva: sac- or cup-like structure at the base of the stem in some agarics.
volval: derived from a volva.

Index

ACKNOWLEDGEMENTS

The author and publishers wish to thank the following for their help in supplying photographs for this book:

Heather Angel: Cover, 4-5, 6-7, 93, 102, 106, 111, 116, 117; Dr Alan Beaumont: 108; Brian Hawkes: 9, 86T, 89, 100, 103, 105; Dr Derek Reid: 82, 86B, 90, 98, 99, 104, 107.

Picture Research by Tracy Rawlings.